PENGUINS ON EVEREST

David Durkan

Dagbladet

Løssalg kr 25,00 € 3 utenfor Norden

Penguin lawyers take legal action against author of *Penguins on Everest* for defamation of character.

> Three penguin lawyers instigate legal action at The International Court of The Hague against the author of Penguins on Everest.
>
> 'Our clients, 50 thousand penguins, are seeking compensation for defamation of character by their being associated with so-called climbers on Mt. Everest.'

Dagbladet - Norwegian national tabloid, not renowned for balanced mountaineering journalism.

An
Invitation

To join me on a lifelong journey from
Wales, via Norway, to Nepal.
Share climbing adventures,
and witness a 'secret' desert war.
Learn of Top Secret NATO plans
to drop nuclear bombs on Norway.

Let us travel together along
The Silk Road
through Turkey, Iran, Afghanistan,
Pakistan, India and Nepal.

Who am I?
A boy from Wales
Father - Grandfather
Traveller - Mountaineer
A lover of life.

•

David Durkan

Kathmandu, 2016

Copyright David Durkan 2012
2nd print/re-write: Copyright David Durkan 2014
3rd print/edited: Copyright David Durkan 2016
Foreword © Doug Scott
Annapurna (Historic Perspective) © Chris Bonington
Annapurna (50 Years of Expeditions) © Reinhold Messner
Murder of the Impossible © Reinhold Messner
The Paradox of Our Age © The XIV Dalai Lama
Only a Person Who Risks is Free © Hugo Prather

David Durkan has asserted his rights under the Copyright, Designs and Patents Act 1988 to be identified as author of this work. The contents may not be reproduced, stored or copied in any form printed, electronic, photocopied, or otherwise, except for excerpts used in review/reference, without the written permission of the author.
This book is based on the experiences and recollections of the author. Quotes used do not imply the said sources share the author's viewpoints. Some names have been changed to respect privacy and to protect the innocent.

The CIP catalogue record for this book is available from the British Library.

ISBN - 9781522799061
Cover and illustrations: Mike Tombs
Layout: Sandesh Prajapati
Printed by: 1st and 2nd Edition - Swami Kailash Outdoor Press
          3rd Edition - Amazon Books
Contact: durkandavid@gmail.com

'The changing face of mountain expeditions has greatly accelerated the last twenty years. Dave's standpoint about today's developments is both critical and constructive.
He asks us to question the threats to the values we have taken for granted.'
Chris Bonington

'As usual, Dave goes beyond the normal conventional barriers. He knows no fear - on or off the rock - to this I can attest.
"Rum Doodle" is the classic of its gene, discreetly disclosing the hidden and shady sides of early Himalayan expeditions. "Penguins" shows no such tolerance, cutting right to the bone.'
Pete Sandall

'David D makes no attempt to take life seriously, yet raises a number of serious points as he sticks pins into sacred cows.
What a delightful read.'
Leo Dickinson

'Warm - exposing - disturbing - very real.
He pulls no punches. Why should he?'
Ian Wall

'As exciting as an Agatha Christie thriller, with tales of vampire penguins competing with killer whales to be the first aquarians to reach the summit of Everest, without eating a Snickers.'
Dave's Grandmother

*There is often greater risk
in not taking risk.*
                    Swami Kailash

*The world is a dangerous place,
not because of those who do evil,
but because of those who
look on and do nothing.*
                    Albert Einstein

Contents

	Page No.
Dedication - To the people of Nepal	VI
Foreword - by Doug Scott	VIII

Book 1 - Under Milk Wood   1
The authors life from childhood in wild Welsh-Wales,
via Ireland, to military service - followed by a journey along
the Silk Road to reach the fabled city of Kathmandu.

Book 2 - Glimpses of Nepal   75
An introduction to trekking and climbing in Nepal.
Places and people.

Book 3 - Annapurna   103
Walking in the footprints of the French, to reach
base camp and witness Barbie's ascent.

Book 4 - Instant Mountaineers   121
Dave, Chris and Reinhold - reflections.

Book 5 - Mount Everest   133
The incomplete story of the world's highest mountain.

Book 6 - Why?   155
Questioning, 'Why we climb?'
The decline of the once noble sport of mountaineering.

Book 7 - Russian-roulette   169
If a Nepali worker dies on Everest -
Is it an accident or is it murder?

Appendix   193
Acknowledgements

Dedicated to the People of Nepal

...and to all who died and suffered on both sides during Nepal's civil war: 1996-2006.

To a nation that lost a generation of 'sons and daughters'.

To a country trapped in poverty, which forced some to join the Army to defend an indefensible royalist-system, while others joined the Maoist insurgency to seek change and improvement. Atrocities were committed by both sides. Over 17,000 dead, 3000 still missing, innumerable people displaced and, as in most civil wars, brother killed brother, and sister killed sister.

Their reasons were much the same:

Some to escape hunger and despair, seeking food, work, education, security and a future. Others were tired of seeing their mothers and sisters die during childbirth on dirty damp floors, and saddened to watch grandparents rotting away in poverty.

All wondering where 50 years of massive foreign aid had gone and why the 'aid' agencies and The World Bank create gigantic debts and leave unsupportable projects? Asking where are the schools, the doctors and the nurses; where is the medicine, the electricity, the clean water and where are the roads?

The answer: *Greed, ignorance and arrogance, egos, nepotism and self-interest have consumed them.*

These questions are just as relevant today as before and during a bloody civil war that traumatised this nation. After a decade of 'Democracy', all calls for justice have been blocked by the very politicians who fostered the war with the promise of national reform and equality. The Nepali Government is rated one of the world's most incompetent, inefficient and corrupt.

Apartheid is widespread and institutionalised through caste and gender. Apartheid is where one section of a nation benefits out of proportion from the work contribution of others in that same society.

All achieved by keeping the people in a perpetual circle of ignorance and poverty, dependency and fear.

Today's despotic regime is supported by a dysfunctional legal arm and a cowered media, which allow the privileged class to plunder the country's resources and siphon off the massive foreign aid for their own benefit - year after year.

The earthquake of 2015 *had long been predicted. The Government went into shock, was paralysed and hampered aid teams, taxing relief materials and confiscating earthquake aid funds. When small NGOs could not pay the high taxes, materials and medicines were confiscated and sold to the highest bidder. Hitler had not stopped the Red Cross, but the Nepali politicians did - with 7000 tents and tarpaulins, 20 trucks and tons of donated and needed supplies all locked behind steel gates* **6 weeks after the earthquake!**

The earthquake was quickly followed by the six-month Indian border blockade, where political leaders slept and people in power manipulated the black market, while the poor paid a terrible price. No fuel, no medicines - yes, people froze and people died.

Almost a year has passed, including a monsoon and a hard winter, and the poor are still living under plastic sheeting. It was and is the police, military, ordinary people (students, young and old, rich and poor) and small organisations who are rebuilding the country - brick by brick, house by house, school by school, while the political parties argue over who gets what of the billions in aid received.

Again, it is time to ask where the yearly billions in foreign aid and newly generated taxes have gone. To ask how these post-war politicians, diplomats, high placed bureaucrats and powerful military and law enforcement officers, and their families, have accumulated so much land and so much wealth in such a short time.

My hope is that one day the corrupt and inefficient politicians, who systematically strangle their own people and country, and the high-maintenance self-serving foreign aid agencies and foreign diplomatic corps, will be replaced by people and systems that place the people of Nepal first.

David Durkan
Kathmandu, 2016

FOREWORD

There is a battle raging between maintaining diversity and losing out to encroaching uniformity. Communities everywhere are struggling to maintain their cultural identity and integrity in the face of globalisation as Mother-nature loses her biodiversity through mankind's arrogance, ignorance and greed.

So too the very soul of climbing and mountaineering is under siege. The trend to manipulate and debase the sport into an activity devoid of uncertainty and risk, devoid of adventure, is growing. Jürg von Känel promoted the concept of 'safe adventure' with evangelical vigour, *'What people want is bolts at a regular distance... set to exclude the risk of injuries in the case of a fall... to enjoy carefree climbing in nature'.*

Such a sport has many elements of urban life: Repetitive monotony and uniformity, safe, predictable, bland and unadventurous.

We only become conscious of our core strengths and weaknesses when making judgements, when being resourceful and creative while accepting responsibility.

Scottish climber Jock Nimlin noted, *'It is by accepting the hills on their own terms... that climbers find their deepest satisfaction... an inner conviction that these principles represent an absolute value... The whole concept of climbing is based on increasing standards of performance with no foreseeable limits.'*

In a wake-up call, Ken Wilson warned of 'The climbing theme park' - and in *Penguins on Everest* Dave sets the alarm by asking, 'What is an expedition and where is mountaineering going?'

He asks us to question if those who struggle up the mountain on pre-fixed ropes to reach pre-prepared camps, established by others who do the work and take the risks, are worthy of being termed mountaineers. Are their ascents of any value? His stand is that instant mountaineers compare themselves to those who first climbed the mountain(s). He notes how they bask in glory because others have died on *their* mountain - while they highlight their own suffering and hardship. They never question their own lack of ability, their own low performance levels, their lack of responsibility nor their own irrelevance.

Dave, *'... to suffer or to die on a mountain does not make one a mountaineer. Nor does one become a hero just because others are suffering and dying on the same mountain. Anyone can suffer, anyone can die.'* He highlights how commercial groups operate under the false banner of national or international expeditions.

How they recruit the unfit and inexperienced to run amok in areas once reserved for those with ability and an understanding of what risk and responsibility entails.

Inexorable bolting on the great Alpine classics results in some routes gaining via-ferrata status. Route finding skills and the consequence of falling are removed from the equation along the bolt line and along the High-Himalaya pre-fixed rope highway.

Retreat should a storm appear, or if the crowds slow them down, or should they collapse of exhaustion, is not a problem: the Sherpa staff assist them down to return home to a hero's welcome.

The safe ascent, up pre-prepared routes done a thousand times before, is null and void by its very irrelevancy and banality.

> *'You never understood that it ain't no good
> You shouldn't let other people get your kicks for you'*
> — Bob Dylan

Bruno Durrer, earlier U.I.A.A. medical commissioner, promotes, *'...that the absolute freedom of climbing consists of choosing to climb wherever you want and with whatever technical help one considers necessary...'* His only criterion is to buy the right equipment and pay the fee.

My stand is that 'absolute freedom' can only be attained when the basic tenets of mountaineering are understood and mastered. Dave, *'...hence mastering the art of moving safely in potentially dangerous places under all conditions is a prerequisite.'*

The mountaineer chooses to face the mountain as he finds her, and not sculpt or downgrade her to suit themselves. Certain practices render the full realisation inherent in mountaineering as impossible to achieve. These include the neurotic search for safe-danger - solved by adding artificial means to alter the conditions presented by nature.

We follow Dave from his childhood in Wales via a roller coaster ride up and down sea cliffs, via a silent war in the Persian Gulf to a frightening behind the scenes look at the NATO nuclear deterrent in his second home, Norway. Here philosopher, Nils Faarlund, influences him to seek the essence of the mountains, *'Speaking of a quality of life, which is archetypal... not related to modernity's shallow fun or high sensation seeking...'* but rooted in the, *'natural rhythms of the planet'.* Faarlund's philosophy challenges us to question the, *'values and lifestyles imposed by modernity'*, and to embrace the, *'joyous satisfaction in the outdoors, an experience of tranquillity'.*

This tranquillity inspires confidence to act in accord with one's personal values, including taking a lonely standpoint (as Dave has done) against powerful and popular easy fix commercially driven mainstream practices.

The devaluating of mountaineering as a noble sport is insignificant in the scale of world events: wars, revolutions, floods, droughts, pollution and economic collapse. Yet mountaineering gives so much at the individual and collective level that its innate values are worth recognising, honouring, practicing and defending.

Penguins on Everest - recommended reading.

Doug Scott

Book 1

UNDER MILK WOOD

'They are the Jarvis Hills,
...which have been from the beginning.'

'Kicking the waves back into the sea...'

Dylan Thomas

The main street in Holyhead, around the time my father met my mother.
Photo: L. Clay

The Durkan Family: *Helena, Dad, Anne, Christopher, Michael, Martin and Mum with baby David. Philip was away. (1949).*

A boy growing up in two worlds:
Protestant Wales: A land of dragons and mad poets
and
Catholic Ireland: Divided by two unforgiving religions.

A child with learning disabilities seeks solace in
The School of the Great Outdoors.
The truth: A paralysing fear of exams.
A teacher's harsh words:
'Join the Army, boy; they use the unusable!'
A climbing Mowgli, on the wet vertical rocks of England,
Scotland and Wales.

Chapter 1

An Irish Catholic in Welsh-Chapel Wales

Picture a 10-year-old Irish Catholic boy growing up in Welsh-Chapel Wales - a land of dragons, wizards, harps and the birthplace of that wonderfully eccentric poet Dylan Thomas.

The farmers and fishermen of pre-Celtic Wales lived in fear of the fierce dragon that ruined their crops and stole their fish. So one day they cast a giant net over the dragon and pulled him out into the sea. In his desperate fight for life, the dragon clawed out giant tracts of earth and stone, which were to form the valleys and the mountains of what we now call North Wales.

Off the main North Wales coast lies the Isle of Anglesey. Off this is yet another smaller island, where we find my childhood town Holyhead. The town nestles comfortably under a rocky mountain, which protects it from the winter storms that rage across the Irish Sea. The Isle of Anglesey boasts a heritage rarely equalled in Britain. It was one of the last places of refuge for the Celtic Druids before they were crushed by the power of Rome.

On this craggy coastline the Romans had their most north-western garrison - perched on the edge of the known world. Here one can still sense the lone sentry silhouetted against a Celtic sky. Anglesey has known peace and war, sun and storms, culture and non-culture, justice and injustice. Saint Patrick sailed along the ragged coastline on his way to Ireland, as did the Norse Vikings. The island was the granary of the Welsh Princes as they fought the invading Normans. Kings, queens and princes have come and gone, as have poets and mad men, murderers and saints.

The Welsh or Gaelic name for Holyhead is Caercybi, meaning Fortress of Cybi. Saint Cybi, was a priest who had inherited the Kingdom of Cornwall; a kingdom he declined so he could devote his life to the service of God. One of the churches he founded lies within the walls of what remains of the Roman fortress.

Here is a gravestone that fascinated me as a child, with an inscription hewn into its cold blue slate:
Remember man as you pass by
As you are now so once was I
As I am now so will you be
And so prepare to follow me.

Today the town of Holyhead is the main car ferry port between the United Kingdom and the Republic of Ireland. In the days of my childhood, that famous train 'The Irish Mail' both ended and started its daily journey here. Tons of letters and brown paper parcels would be transferred to the *Mail Boat*, and after midnight this hardy ship would plough its way through the cold black Irish Sea to reach Dublin at dawn. So too, the train would speed its way back towards London to deliver the mail from Ireland. One could almost hear its iron wheels singing along the steel rails as the train passed through the sleeping countryside.

It was on this very same train that countless young men and women embarked upon their journey to the land of opportunity, America. My father, I used to believe, had crossed from Ireland on his way to that fabled land. The story goes that he got off the *Mail Boat* and, instead of taking the train to London, he had walked straight into the first pub he saw. Here there was a roaring coal fire and Guinness* on tap - and warm Irish voices that wished him welcome. So he remained in Holyhead, married my mother, and sired seven children.

Eire - For many of the Irish, emigration to America was the road to economic salvation. Their country had long been ruled by the ironclad fist of the British Crown, subjected to wars, deprivation, poverty and unemployment. Being a bastion of Catholicism did not help either, as large families meant lots of mouths to feed. As for the emancipation of the Irish women, that had to wait a while. Any form of birth control other than the natural fertility cycle was condemned by the Catholic Church; as such contraception was unavailable.

Should a young girl become pregnant outside the bonds of matrimony she had four options. One was to marry the father of the child-to-be, an option he would usually decline, as she was *a tainted woman*. Her second alternative was to give birth and bear the stigma of having a child without a man - with the isolation and poverty that that entailed. Or thirdly, to give birth, only to see her child placed in a convent orphanage or adopted.

* **Guinness.** *'Ireland's national drink': Dark 'stout' or 'ale' produced using water with a high mineral content, top-fermenting yeast and black-roasted barley malt.*

The fourth alternative was abortion, something no one talked about as it was illegal in Ireland. The pregnant girl could take the *Mail Boat* to Holyhead and continue her solitary journey by train to Liverpool. Here in hospital she would lie with patients who carried the diseases of the poor and the dying: tuberculosis, diphtheria and whooping cough. Then, with the shame removed from her body, she could secure work in a factory, or in the gin-shops and whorehouses of the city. Some girls would return after a few years, their sins forgiven and forgotten. They received a second chance, that of looking after their ageing parents, or marrying a fine man. Yet such men were often twice their age, unemployed, drunkards, or possibly retarded - a fine menu to choose from.

The men themselves were faithful to the Catholic Church, for the priests turned a blind eye to their beer and gambling. Any sins, such as the beating of wives and children, would be forgiven in the Confessional Box. Poverty breeds an unquestioning and uneducated flock of the faithful, who become addicted to the doggerel, pomp and ceremony of organised religion. This control of the masses by Church and State is not just a Catholic phenomenon, but is practiced world-wide. Religion gives men a false sense of their superiority and importance, while keeping the woman pregnant and in a state of dependency and fear.

The first born son would normally get the best education the family could afford. The second son would often work on the small family farm, or become a manual labourer. The third son would travel to Dublin or Belfast to find work in a factory and marry the first girl that would have him. Any boy who came later would have to compete for both food and love. Emigration was one road or he could seek solitary comfort in a beer glass. Girls helped around the house, were cheap to feed and schooling was not important, for they would marry young and bear seven children of their own.

In my father's case, it was not the story of gloom and doom that I had believed. He had not come across from Ireland and gone straight into the nearest pub. His family had emigrated to Holyhead from County Mayo, in the west of Ireland, many years earlier. Here they had found a thriving Irish community. Dad worked at The Cattle Yard, where cattle from Ireland were bedded and fed before continuing their final journey to the slaughterhouses of England.

Love Story - How my Irish-Catholic Dad, Joe Durkan, met my Welsh-Chapel mother, Lilly Clay, I do not know, but we do know that she was from a higher social and economic class. Worst of all, her strict Protestant background was governed by threats of eternal fire and brimstone. The young Joe and the even younger Lilly must have fallen helter-skelter head-over-heels in love and one can only imagine the trials and tribulations they faced. Her parents disowned her, while the Catholic community were none too happy either.

So Mum and Dad ran away to the seaside town of Rhyl. Here they lived in a bed & breakfast, while Mum took lessons in Catholicism. She had to sign an agreement to bring up all their children as Catholics before the priest would sanctify their union.

She proved to be fruitful, and the birth of one child followed another in good Catholic tradition until there were seven siblings. For my mother a deal was a deal, and we children paid a heavy price. We attended confession every Saturday, mass with communion every Sunday and we never missed Sunday School. I could hardly read and write at school, but I could recite the liturgy of the Mass in both Latin and English, word perfect.

The clouds of war - When the Second World War broke out Mum, Dad, five children and a dog were all living within walking distance of St Mary's Church, St Mary's Catholic Primary School and The Catholic Club. The latter played an important cultural role, as after Sunday Mass the faithful went straight to The Club for a talk with the Parish Priest - who also liked his pint of Sunday Guinness.

With war, in 1939, the country's young men went to meet the threat of Hitler's Third Reich. Since Dad brought cattle over from Ireland, he was a part of 'The Food Chain' and as such he was never conscripted into the Armed Forces. Within a year of the war starting, many young men joined the military, so there was soon a lack of policemen and firemen in the town. To help resolve this, Dad was given a policeman's jacket and helmet and with no training, was told to look out for suspicious people coming off the *Mail Boats* from Ireland.

There was no passport control between the United Kingdom and Ireland. Hitler could have invaded Britain by buying tickets on the Irish Ferry. Yet he was probably deterred by mild-mannered Joe Durkan, father of five, sitting there in another man's policeman's jacket - truncheon, whistle and all.

When all the passengers were through the gates, Dad took off the policeman's jacket, donned his farmer's clothing and quietly cleaned, feed and bedded down his cattle. Yes, he helped feed the nation while the country was at war, something he was silently proud of.

Just after the war ended Chris was born, and I came along in 1949 - we were now seven children. I was neither planned nor particularly welcomed into this poor family. Luckily, my birth was about the time my mother's father died and her mother, Granny, had become frail and forgetful. My mother's sister and two brothers had moved away from Holyhead, so they could not look after their infirm mother. One solution was our family moving into the family house to look after Granny. So a deal was struck. Mum took over ownership of the family house lock, stock and barrel.

You can imagine the excitement and chaos as a family of ten: Mum and Dad, five boys, two girls and a long-legged scruffy dog called Sandy moved into what was a virtual palace.

With just Dad's wage to feed and clothe all seven children, we were an economically deprived family. Always clean and well fed, yet we had to use hand-me down clothes: jumpers with repaired elbows, worn out trousers and scruffy school jackets. Mum, now relieved from the chore of being pregnant all the time, turned the downstairs front living-room into a café. This she opened three days a week, specialising in morning coffee and afternoon teas with homemade scones and cakes.

So that Mum could run the 'tea room', my two sisters, Anne and Helena, were given the job of babysitting me. This they did by locking me in a small dark cupboard for two to three hours a day so they could go and play!

Both sisters deny this.

Papists and Prods - A Papist is a Catholic, many of whom were recruited into The Irish Republican Army, IRA, who had fought a long guerrilla war to free Ireland from English oppression. Yet the country was still not united, because the people of Northern Ireland were mainly Protestants, known as Prods, who did not wish to be 'liberated'. Intense fighting continued in the streets and in the countryside of Northern Ireland. Bombings and executions were weekly occurrences, resulting in cities like Belfast being divided by barbed wire fences.

About the time I reached the age of seven, Mum was admitted into hospital, which was followed by a long convalescence. This meant she had problems coping with seven children and a café. To lighten her load, Chris and I were sent to live with relatives. I went to Auntie Jane and Uncle Eddy in the seaside village of Dalkey, just outside Dublin. Here I soon made friends with the boys next door; they were rough and tough types with holes in their jumpers, patched pants and no socks in their shoes. We fished off the pier and we rowed across the straights to Dalkey Island in a leaky wooden boat.

In Wales, we had played cowboys and Indians; here we played Papists and Prods. We Catholic children were the Papists and would shoot the dirty, stinking Prods and they would shoot us. Unfortunately there were no Protestant children in the village, so some of us had to pretend to be dirty Prods.

Oops! I nearly forgot; there was one Protestant family in the village. The Cassidy family, a mother and father and two boys, about whom my new friend Patrick proclaimed, *'We Catholics don't play with dirty stinking Prods!'*

One day I met the younger of the two brothers as he was kicking a football about on his own. I automatically tackled him and we were soon passing the ball back and forth as we raced down the street. Potential world champions both of us, when to our horror we watched as it rolled down the side path and into the sea!

'Shit!' he said. *'That's my brother's new football; he'll kick buckets of green shit out of me.'*

I ran into the reception of The Castle Hotel and snatched one of the fishing rods their guests used and we soon had the ball back on dry land. We sat on the pier and shared our plans and dreams. I was going to play rugby for Wales and he would be the Lightweight Boxing Champion of Ireland.

I asked him what he was going to wear to the Easter Festival and, before he could answer, I told him I did not have a fancy dress and shared my deepest secret, *'I am frightened of water, well not frightened of having a bath, but of the sea, as I cannot swim.'* He told me his name was Joseph and that his mother was totally mad. She collected Swiss cuckoo clocks that went cuckoo at the same time, which drove his Dad absolutely crazy.

The next morning Joseph's mad mother called at our house and left a Red Indian dress with real feathers and a rubber tomahawk. She said Joseph had grown out of it, so I could use it for the Easter Day Parade. Wow! I would be a real Red Indian - just like the ones on TV.

Unfortunately Patrick found out that I had talked to a dirty stinking Prod and he attacked me like a mad man. I felt so alone standing there, too frightened to fight back, my eyes filled with water and snot bubbled out of my nose. I did not cry.

The Village Festival - Saturday came and with real war paint I was Black Feather, War Chief of the dreaded Apache Indians. Patrick and I were best friends again and he was dressed as a fierce pirate, with a plastic sword, eye patch and a stuffed parrot on his shoulder. In the smoke-filled bar the men were downing glasses of Guinness; the level of talk and singing grew louder and louder as the evening drew on.

Soon bored, Patrick and I discovered six pairs of Wellington boots in the hotel's entrance. When the receptionist was not looking, we half-filled them with water and with gleeful delight could picture the shock as guests put on their wellie boots! Ten minutes later we ate red wobbly jelly and coconut buns and I danced a war dance. I looked around for my newfound friend Joseph and his brother, but they were nowhere to be seen.

One God - Of all the children in the village, it was only the Cassidy brothers who had not attended the Easter carnival. They had been excluded, a situation that was 'understood' by both sides. In Wales I was a Catholic and my school friends were Welsh-Chapel or Church of England. Different churches and different hymns, but we worshipped the same God.

In Ireland, on the other hand, I learnt to really hate Protestants. One children's chant I remember well:

Ahem, ahem, my mother has gone to church
She told me not to play with you, because you're in the dirt
It isn't because you're dirty; it isn't because you're clean
It's because you are a Protestant and eat margarine.

The logic was clear: butter is better than margarine and Catholics are better than Protestants. Ireland has come a long way since then; there are two child families and peace has been declared in the North. Yet they still sing songs of death and destruction, of war and hate, in the pubs on a Friday night.

Durkan's Café - While I was away in Ireland for three long years, Mum had turned the café into a restaurant, serving lunches and dinners. She did the cooking, while my sisters served at the tables and did the washing up. She opened a tobacco and sweet shop and my brother Philip built an outside ice cream parlour. Here I worked after school and at weekends, a schoolboy's dream! Mum then bought the house next door, a proper home into which the whole family, dog as well, moved.

Now the business was ready for expansion. Upstairs, in what had been granddad's and grandma's bedroom, Filip built a coffee bar with a steam driven coffee machine. Mum rented a jukebox, with 50 of the latest records: Elvis Presley, Pat Bone and later, the sinful Rolling Stones! Within weeks, the teenagers of the town flocked to the new Durk's Café every Friday and Saturday evening. They bought their Cokes, milk coffees, ate their KitKat and Penguin biscuits and listened to the songs of Patsy Cline and Jim Reeves. The talk was about exams, holidays and who was dating whom, and there were endless tales of broken hearts. Durk's Café became the 'in' meeting place, shaping the dating, mating and family constellations of Holyhead for three generations.

My disdain for organised religion had taken root by the age of 12.

Chapter 2

Holyhead County Secondary School

At school, my exam results were dismal and I, together with Tommy Lloyd, was always causing disturbances in class. When caught doing something wrong we were sent to the Headmaster to be caned. Caning was a primitive method of punishment where a grown man takes a finger thin bamboo cane and, with all his might, strikes the hand of a small boy. Not once, but usually five or ten times.

In addition, I was not particularly good at football, feared rugby, could not swim and was bored by cricket. Sailing and golf did not appeal, nor had I any talent for painting or for playing a musical instrument. As for singing, the music teacher ordered me to mime. I was useless at woodwork and mechanics. I also had difficulty formulating written answers in class. It was, therefore, assumed that I was rebellious, lazy or stupid - or all three.

Being the class idiot was a lonely place to be.

About the same time, I developed a stirring interest for those strangely shaped creatures we call girls. I fantasised about them and wondered if I was going mad. A parallel development was discovering the works of Welsh poet Dylan Thomas, whose words I devoured. Reading *Under Milk Wood* under the bedcovers, I would fade into dreamland. However, poetry was for girls and queers (homosexuals), so Thomas became my secret of shame. I read and lived his works along with Coleridge's *Xanadu* and *The Ancient Mariner.*

Unfortunately, a low academic achiever with a face full of teenage spots, with no sporting achievements and who read poetry, was not attractive to schoolgirls with bumps on their chests.

Mountain classroom - There was one option left and that was joining the school hill walking club, which was organised and led by Miss Hope, the school's science teacher. We used to stay in her sister's small cottage on the slopes of a lonely hilltop. One weekend led to another and we were soon walking and exploring the rolling hills of North Wales. In addition, some of the girls had shapes like Venus de Milo, with the advantage of having arms. Miss Hope was open and warm and genuinely enjoyed our company. Then one weekend, totally to our surprise, she took us to an old farmhouse she had just bought. Unfortunately, it was not just old, it was seriously dilapidated. There were holes in the roof, broken windows and no front door, resulting in the floors being ankle deep in layers of sheep shit.

We would drive there each Friday after school and put up our tents in the adjoining field - rain or shine. Within two weekends the house and barn were sheep-shit-free zones. We pulled down walls and built new ones, scraped and painted and chased out the rats. Ditches and drains were dug and trees chopped down for firewood. We cut fingers, bruised our knees and blistered our hands, but there was never a complaint.

Original cottage, with milkmaid.   *New cottage, after re-furbishing.*

In the evening Miss Hope and the girls prepared a giant dinner of Welsh farmhouse stew. Then we would sit around an open coal fire telling ghost stories.

Sunday was our mountain day and we went hill walking right after breakfast. Sometimes we walked in bright sunshine and other times in the rain and mist. As winter arrived, we would tramp the hillsides in the wind and snow, wrapped up in woollen jumpers and oversized Ventile anoraks.

Miss Hope taught us how to navigate by reading the lie of the land. We learnt to remember the noise of bubbling streams or the feel of marshy patches under our feet, so we could trace our way back in the mist. We knew which way the prevailing wind blew and navigated by the direction the trees bent. By the formation of the clouds and by the colour of the sky we predicted the weather. She taught us how to sense our way through the mountain landscape.

One Friday, we were a large group going to celebrate the completion of our work on the cottage when unexpectedly her car and the minibus took a detour. Thirty minutes later, to our amazement we ended up in the Ogwen Cottage Mountaineering School car park.

Ogwen Cottage and Tryfan.

In total secrecy, Miss Hope had arranged and paid for a weekend's climbing course for all of us. We were going climbing, with ropes and helmets, just like real mountaineers! Sixteen excited school kids did not sleep easy that night!

Each move led to a new challenge and all around was rock and more rock - it was a gigantic Mad Hatter playground. I became Edmund Hillary and Dr. Livingstone rolled into one. My face was alight and the tips of my fingers burnt as they searched for handholds. Here, there were no exams to frighten or paralyse me; here there was only one examiner: The Rock.

So on Monday, I sold my 3-gear bike and Disney figure Timex watch. Then after school, I hitchhiked thirty miles to the village of Bethesda to buy a climbing rope, one sling and one steel carabineer. The next evening Pete Jones and I walked to Holyhead Mountain and climbed two routes. One we christened *Hat* because Pete's hat blew off, the other we named *Plimsoll,* after the gym shoes we climbed in. Soon other climbs followed. There was *Wally's Folly,* because Robert,

Climbing Tempest and The Pealer - First ascents - Holyhead Mountain. (1966/67)

whom we called Wally, went the wrong way and almost fell off. There was *Curtains,* because if you fell you would probably break your back: curtains for you! Followed by *Tension, Teaser, Tempest, Patience, Cursing* and *Bloody Fingers* - all names that reflected the character of the route or the experience we had when first climbing them.

By today's standards, these climbs are not high in the scale of difficulty, Severe to Very Severe (grades 4 and 5), yet, the climbing quality was excellent. We fourteen-year-olds pioneered over forty new climbs without an instructor and with almost no equipment. It was more through luck than skill that no one was injured or killed.

We then moved to the mountains of Snowdonia and climbed virtually every weekend, telling our parents we were going hill walking. We slept in barns, in the back of cars and camped under giant boulders. When not climbing, we would spend our time reading mountaineering books and pondering on the mystery of girls.

First book - *Annapurna* was a riveting story about the first successful ascent of an 8000m peak in Nepal. Then came *Seven Years in Tibet,* which evoked a hunger for travel and more books. Buhl, Bonatti and Terray, mountaineers all, took the place of Dan Dare, Superman and Batman. Climbing and adventure travel reading helped my schoolwork considerably, yet, I was still behind the rest of the class. One teacher, Mr. Bacon, in a fit of rage at my seemingly never ending stupidity, thundered: *'Join the Army, boy. They use the unusable!'* Not particularly warming words, but he did have a point.

First date - Wally Owen was probably the first of our crowd to 'go all the way with a girl', an exploit that increased his status to group patriarch. So, when in doubt we sought his advice, *'When one is as ugly as you, Durkan, with no natural charm, take her to the cinema. The darkness might hide your handicaps!'*

I took his advice and, to my surprise, the girl of my dreams, Mary (name changed), said, *'Yes!'* We went to the Empire Cinema and sat in the back row watching *Smiley Gets a Gun.* Halfway through the film I put my arm around her shoulder, no protest! She even snuggled closer so I slowly started to tentatively touch her left breast, again no protest, so I progressed to the right one. Soon, I was massaging both her firm pointed breasts with great enthusiasm uninterrupted for the rest of the film.

When the lights came on, I discovered that Mary had crossed her arms over her chest and I had been furiously rubbing both her firm pointed elbows. She gave me a polite *'No'* to a second date.

Chapter 3

Big sister

At the age of fifteen I took my CSE-exams and the results were dismal. The choice was to stay back one year and re-take the exams or leave school and find a poorly paid menial job, or take up a life of crime. I chose a fourth route and ran away from home.

Our Agricultural Science teacher 'Dig' had allowed Tommy and me to feed the hens and do small gardening jobs to keep us out of trouble. Hence Agricultural Science became my favourite school subject. Dig had told us about a youth apprenticeship scheme in Horticulture and Forestry. Here they accepted 16-year-olds as trainees, so, in total secrecy, I filled out the forms, changed my date of birth, forged my father's signature and popped the application in a post box.

A month later, I was working at Vaynol Estate, under Snowdon, which was a great woodland park belonging to Sir Charles Duff. Here they provided work clothes, a bed, two hot meals a day, a sandwich pack, plus a symbolic wage. Weekdays were spent planting tree saplings, weeding, cutting down brush and going to evening classes twice a week.

Weekends were devoted to climbing the classic rock routes like *Faith* and *Hope*, *Pinnacle Ridge* or *Soapgut* in the Snowdonia National Park. Often, I simply went walking or climbing on my own. Forestry was backbreaking, soul-destroying and poorly paid, so after three months, I packed my bags, collected my wage and hitchhiked to my sister Anne's house in Chester. Six hours later I turned up on her doorstep with a rucksack of climbing gear and a plastic bag full of dirty washing. Anne opened the door and looked at me standing there dripping wet. She didn't blink an eyelid.

'Come in. Take off your shoes, so you don't dirty the new carpet.'

After dinner, Anne phoned Mum to say that I was safe. Mum had called the police to report me missing, but there was little they could do. A mother of seven knows about life and she had probably assumed I would sort myself out or kill myself in the process.

In Chester, Anne found me a job in a health food shop called Dutton's, where I sold seaweed and bird-seed cakes to elderly ladies seeking eternal life. After a few weeks, Mr Dutton offered me a three-month evening course in Junior Managerial Training, and noted, *'If successful, you may one day manage the Health Food Department.'* Thoughts of a future promotion with subsidised canteen meals, free uniform jacket and shop tie, fourteen days paid annual holiday with a pension! What an opportunity for a 16 year old who had just failed his school exams!

Thanks, Mr. Dutton, but no thanks. I was 16, no longer a virgin and wanted bright lights and fast living. In addition, I could not live forever with Anne and her growing family. She was great, but we realised I needed some form of direction in life. At which point, my teacher's harsh words came thundering back:

'Join the Army, boy. They use the unusable!'

My new mother - I decided to join the Royal Air Force (R.A.F.), who accepted recruits at 16 years of age for an apprenticeship scheme. Unfortunately, being under 18 years old meant I was legally a minor, which necessitated my parents' signed permission. Not wanting to ask Mum, who was hardly talking to me, or ask Dad, I persuaded big sister Anne to pretend that she was my mother.

She was game, but there was only twelve years age difference between us. This discrepancy we solved by putting some socks in her bra and dressing her in an old fashioned mother-looking type coat, adding face powder and thick lipstick - which all helped the ageing process.

The Recruiting Sergeant shook my new 'Mum's' hand, hardly casting a glance at her. That was a relief. He explained that basic military training would be followed by two years general educational and technical training in a useful trade. Then, I could serve for twenty years and get a full military pension!

Anne smiled at him, knowing that a pension was the last thing on her little brother's mind. She took a deep breath, having passed the point of no return, and signed on the dotted line:

Mrs Lilly Durkan

We walked out of the office absolutely petrified that they would discover the truth and we would end up in some military prison to be placed against a wall and shot at dawn.*

---

* Sister Anne claims she never dressed up as our mother.

Chapter 4

Boy soldier

I became a boy soldier and joined one hundred other sixteen to eighteen year olds on a twelve-week basic training course, a time of torture and sheer terror. We camped in the rain with bad equipment and marched at midnight, with heavy packs and wearing uncomfortable boots. We went to the gym, studied military history, learned how to survive nuclear fallout and how to extinguish crashed aeroplanes. There was training in the use of firearms and in hand to hand combat. We learned how to catch, kill and cook a hedgehog, should we be stranded after a plane crash. *'Anyone for hedgehog and squirrel pie, acorns and dandelion salad?'*

Our instructor, Sergeant Broadhurst, seemed to hate us and spent the next twelve weeks determined to break us, body, mind and soul. At 5 o'clock in the morning he would enter our sixteen man room and throw a bucket of gravel onto our newly hand polished floor, screaming at the top of his satanic voice, *'Hands off cocks, on with socks, you horrible little men!'* Then off we went for a one hour run in the dark, cold November rain.

Sgt. Broadhurst immediately picked out Reginald (Regi) Farmington, William Williams and I as his personal targets of hate. Regi, no doubt, owing to his irritating upper class accent, and Williams because he had an obvious feminine disposition. Why me? Probably because my slight frame indicated physical, mental and character weakness, something military training was designed to filter out. *(Photo: Me at 16, boy soldier).*

Williams was a bit weird and a loner, whereas, Regi and I were survivors by nature and we immediately joined forces. We refused to bend to Broadhurst's satanic will, which only added fuel to his fire. It became a battle of will and determination between Sergeant Rasputin, an inhumane sadistic bastard, and us, two innocents. He pushed and pushed our physical and mental limits to the edge, humiliating us time and time again. Some mornings I awoke in a sweat, teeth grinding and muscles screaming. Regi suffered less, guided by a very non-Buddhist mantra, *'Screw the bastard!'*

William Williams he pushed just as hard. One late afternoon we returned from the firing range to find his bed stripped and all his personal effects gone. The pressure had been too much for him. They had discovered him sobbing like a child, wrists slashed, sitting in his own blood and urine. Williams was discharged on medical grounds.

1 - 0 to Sgt. Broadhurst - This may well have been Broadhurst's job, to weevil out the weak links, but it really pissed us off. There had to be a response. Five days later, Sgt. Broadhurst's pride and glory, his pale blue Morris Minor car, sank to the bottom of a small lake near R.A.F. Cosford, never to be seen again. The score was now 1 - 1.

After basic training, where Regi and I ranked in the top ten cadets, a series of tests followed to see where our real talents lay for further vocational training. Regi went to Officer College and became a fighter pilot. My results indicated that I could spell my name correctly and had 'precise coordination skills'. This meant I had the potential to be a radar technician, explosive demolition specialist, or a brain surgeon! But alas, I lacked any grasp of maths, algebra, physics or the sciences.

Suddenly, I was on a three-month Communications Specialist course, training to be a telex operator!

Adjusting to military life was not easy, my strong allergy to authority being an obvious handicap. We Junior Aircraftsmen were the lowest rank. We had classes and training five days a week and at weekends we were confined to the barracks. One needed a valid medical or humanitarian reason to gain a 48-hour pass. I could always have my parents murdered to attend their funeral, and then go climbing.

Not being able to understand the logic in sitting on my backside every weekend, the solution was obvious: ignore the rules. Observing that the guards did not normally check those going out to the village pub, I formulated a plan. Each Friday, I simply walked past the guard room as if I was on my way to the pub. Here I had a half-pint of cider and then exited via the back door and, with my thumb raised, headed towards the mountains of North Wales. A rucksack and a rope across my shoulder normally meant a climber would pick me up, to arrive at Cobden's Pub in Capel Curig, before closing time.

Here I would meet friends, or if necessary pick up a stranger, and go climbing the next day. Sunday usually saw me going for a climb on my own before hitchhiking back to the camp. On reaching the village, I popped into the pub before returning through the Guard Room gates.

This ploy worked three weekends in a row, but alas, on the fourth Sunday, a guard stopped me. He insisted that I open my larger than necessary to go to the local pub shoulder bag. This was followed by having to explain why I needed a 120ft climbing rope and a sleeping bag. Tricky? Yes, an impossible task, resulting in my being declared AWOL (**A**bsent **W**ith**o**ut **L**eave).

The immediate consequence of being AWOL was being locked in a cell for the night! The punishment, next day, was quick and painful: continue lessons during the day, scrub floors during the evening and spend seven nights behind bars. In addition, I had to stand in a large metal dustbin at lunch time and salute any officer that walked past, shouting, *'Sir, I am a dustbin full of rubbish, Sir!'*

After training I was transferred to the small village of Linton-on-Ouse, just outside the city of York, with the exalted rank of 'Leading Aircraftsman'. Here trainee pilots learnt to fly Jet Provost aircraft, and my job was to type their flight plans. Boring, boring and boring...

At Linton discipline was far less rigid, I could go climbing every weekend and study in the evenings. Things were certainly looking up. Together with Porky Smith, I started a Folk Music Club, which led to my first real love, Maggie Cotswoll (name changed).

Oops, I forgot to mention my sexual debut. This happened some months after the cinema elbow rubbing fiasco, and occurred behind the Holyhead Park tennis courts. It, like the cinema date, was nothing to be proud of. To put it mildly, it was fumblingly terminated before it really got anywhere. She, for her part, did not ridicule me, but with great understanding coaxed me along the lines of: *If at first you don't succeed, try try again.*

This act of kindness, and her further willingness to give lessons, did reveal that certain girls have a fascination for the state of virginity. They, usually slightly older, see the assisting in the de-flowering of young men as vocational - even a form of sport.

On realising this, and insisting that they be gentle with me, I sacrificed myself and lost my virginity on five different occasions over the next eighteen months.

Chapter 5

The Force of Gravity

Climbing was the first priority - with work and girls poor seconds. Some weekends, we would burn rubber and head to the Lake District in Bob Williams' little Mini Minor. Bob provided the transport and I taught him to climb. I was young, physically fit, gung-ho and took risks, while Bob, the enthusiastic novice, was under the impression that I knew what I was doing. We slept in his car outside a pub in the village of Borrowdale and the landlord gave us free dinner and beers in return for Bob singing songs of love and war. His sick party piece was *'Teenage Cremation'*, a heartrending love story about a biker who lost his true love in a confrontation with a lorry:

> Ninety miles an hour, we bumped into a lorry
> The driver he got down, he said 'I'm very sorry.'
> Now from Watford to Bradford, Peterborough too
> I walk along the side of the road, picking up bits of you...
> Refrain: Teenage Cremation - oh how I cried when you fried!
> I sit here by the furnace; I listen to you fry...
> I hold in my left hand, your ear and your eye.
> But they aren't the pieces I wanted of you
> They lay on the road between Peterborough and Crew...
> Refrain: Teenage Cremation - oh how I cried, when you fried!

Other weekends were spent in Scotland, getting lost, in the Peak District, climbing on the gritstone crags, or in the mountains of North Wales. Here we struck lucky. One of Britain's leading rock climbers, Pete Crew, was exploring a secret cliff of giant rock walls that rose like Titans out of the Irish Sea. The elusive Pete was writing a small guide book to the cliffs, and he generously drew us a diagram of the climbs that had been climbed - a real prize!

Our equipment was still pretty basic and we had not climbed at this level of seriousness before. Yet, armed with Pete's diagrams we climbed *Shag Rock, Puffin* and *Panting, Dirtigo, Bloody Chimney, Simulator, Bezel, Ramp* and *Gauntlet* and met up with more experienced climbers. Under their guiding wing, I climbed the classics like *Gogarth* with Lou Costello, and the amazing *Mousetrap* with two members of the infamous Rock and Ice Club.

Both A, Simulator, and B, Imitator, lead to C, Gordon Bleu, and D, Diogenes. 40 years ago, I solo climbed these routes high above the Irish Sea.

I loved the smell of the sea and the sound of the waves crashing below. The metamorphism of boy to climber was underway.

I fell 60 feet once - Sort of flew off *Yob Route,* a climb in North Wales. Sixty feet equates to 20m and it was a free fall, which meant touching nothing until I landed. Yet, Lady Luck smiled as I crash landed in a muddy pool between two rocks and went right up to my knees in thick mud. This soft landing no doubt saved my life.

I remained unconscious for two to three days and, when I did wake, I thought I was dead. All around me were giant fairies, moons, twinkling stars and smiling suns, flowers and dancing milk maids with giant bosoms, happy cows and flocks of fluffy sheep.

Was this heaven or was this hell? No, it was the annex to the children's operating theatre in a small country hospital, hence the dancing ducks and happy hens. My lower spine was compressed in three places and the upper neck vertebrae were distressed, whatever that meant. Ribs were bruised and lungs were in trauma. My face had hit my knee and the whole right hand side had imploded, eye and all, plus my nose was flat as a pancake and lay to one side.

It was as if someone had hit me in the face with a frying pan.

I struck gold again. One of Britain's leading surgeons had chosen to live in North Wales because of the excellent trout fishing. Another stroke of luck was his friend, Sir Bumbly-wobbly-something, was on a fishing holiday. This sidekick was Queen Elizabeth's personal surgeon and between fishing trips these wizards of the scalpel sandwiched in three operations. They reconstructed my face, placing my eye where it belonged and repositioned my nose.

Parallel to this, another medical team constructed a Lego-like traction device to pull and release my legs and joints. They used da Vinci-type lifts to reduce the pressure on the lower spine and pelvis, while I lay moaning and groaning on wooden boards.

Another doctor kept sticking needles in my toes and legs to see if I could feel anything. Bloody right I could, '... *you f=c%&ng-pervert - I know where you live - I'll eat your daughter's rabbit alive!!*'

In the end, he grudgingly accepted that there was full nerve connection between head and feet. This was confirmed each morning when a nurse rubbed my back and legs with oil to reduce bed sores, an exercise that resulted in embarrassing erections. On the bad side, Maggie had dropped me as her boyfriend, but as compensation she had knitted a woolly jumper to keep me warm!

On the lighter side, I fell in love with a trainee nurse and her with me. She read aloud mountaineering books by the side of my bed when off duty and solved the noted embarrassment. Even after all these years, I can still see her face clearly, see her smile, feel her hand movement, and above all, hear her voice reading Joe Brown's *The Hard Years*, followed by Bonington's *I Chose to Climb*.

After three months in hospital, they sent me off to a convalescent home, a sad parting between patient and nurse.

At the new place, I was subjected to five days a week of torture on a apparatus from The Inquisition. My physiotherapist and inquisitor virtually stepped out of a pin-up magazine, except she had her clothes on. Her first action was to remove my spine support corset, open the window and say, *'Let's get rid of this contraption.'*

She was a child-eating zombie and her marble-glazed eyes were the stuff of nightmares. She pushed me to the edge of the planet - yes: It is flat! Then the seven-headed monster would bring me back, let me rest, only to repeat the process. I walked out of rehab after six weeks and was back rock climbing five days later.

I still send her the odd nod of sincere gratitude.

THE KING'S SHILLING

Maldivian baggala. Acrylic on canvas, 2009. Artist: Xavier Romero-Frias.

In the 18th century soldiers and sailors were paid in the
region of 1 shilling a month
(food, uniform etc being deducted).
On 'taking the King's Shilling'
they committed themselves to serve 'unquestioningly',
King and Country.

Chapter 6

Sun, sand and blue seas

Two weeks after coming out of hospital, a Royal Air Force doctor declared me: '*A1 physically fit for active duty overseas.*' Ten days later, I, together with two hundred other 'cannon fodder' low rankers, was on a troop transport flying to Bahrain, a little island in the Persian Gulf. Lady luck smiled again: sun, sand and blue seas.

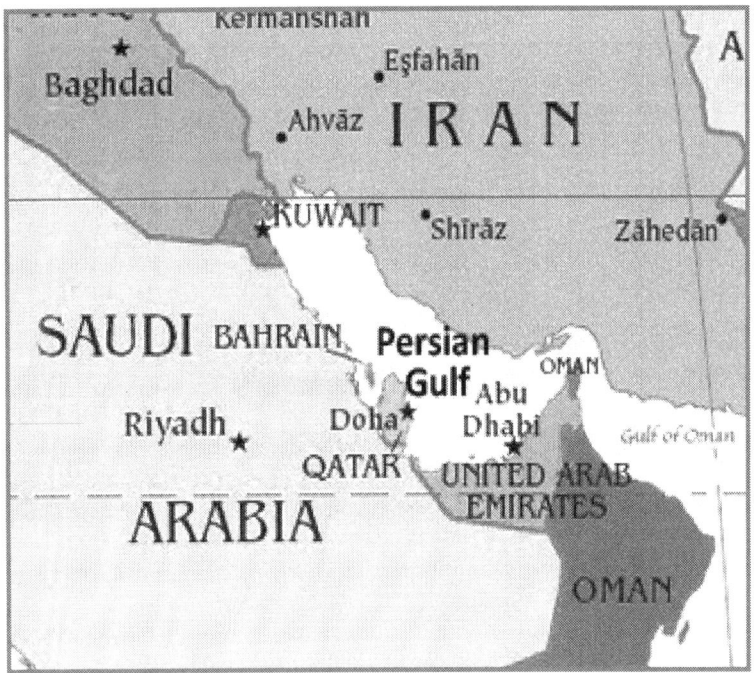

Here, we worked long and very boring shifts in a communication centre, sending and receiving messages to and from military forces all over the world. In our free time, we watched John Wayne films and drank ice cold beer in a male-only environment. The week's main entertainment was often a fight in the bar between Army and Navy personnel, where a broken beer bottle re-structured the face of the opponent. A skirmish that could easily turn into a team event, with everyone throwing punches.

Another game we played, where heavy bets were placed, was *The Dance of the Fiery Arsehole*. Here, two teams each chose a representative, usually a volunteer. He then stripped naked and stuck one end of a toilet roll between the cheeks of his posterior. This was 30 wipes long, and at the start line, the other end was set alight. The two 'athletes' then ran across the bar room, or outside in the parking area. The winner would be the one who ran the furtherest and had the least wipes left, allowing the fire to come as close to his posterior as possible - before releasing it or it having being extinguished.

If that did not light your fire, you could take a cultural trip to the local whorehouse. This was comparable to sticking your dick in a wasp's nest, with similar results: a penis covered in warts and boils!

Others drank themselves into regular oblivion.

We were working class soldiers doomed to remain in the lower ranks of society, and be paid low wages. To increase their income some of the men formed blood teams of six to eight members. Each month, one of the teams went to the local Arab Medical Centre to sell blood. Here they received a cup of tea, some digestive biscuits and US$500 cash each for a pint of blood. Half of this went to the donor and the other half into the Blood-team's beer kitty.

I decided that there was no future in bar brawls, whorehouses, or in selling blood to earn beer money. There had to be a better way to pass one's time. There was; the island of Bahrain had a number of rocky outcrops just waiting to be climbed.

Only ten people are necessary to start an official military sports club. Within a week, I had recruited enough members to form 'The Bahrain Island Mountaineering and Sailing Club'. With official sanction, we gained time off work, the use of a Land Rover and a permit to explore the island. We filled the jeep with boxes of iced Tiger Beer and tubs of fried chicken and off we went on a journey of exploration.

The island's crags offered short yet steep climbing on near-perfect hard sandstone, while the vultures followed our progress with interest.

In addition, we bought a second-hand canoe and six of us conducted a midnight commando raid against The Royal Navy Sailing Club. Here we stole a JP-yacht, re-painted it, and sailed out into the Gulf waters, learning navigation as we went along.

Me leading two new climbs: 'Scorpion Wall', V, and 'Sand in my Sandwiches', V+.

We met and followed the wooden dhows as they ploughed the ancient Arab trade routes. We canoed the shallow waters around the island's fishing villages, skimming over leopard rays as we darted between saltwater reeds. The fishermen waved us in as they sat repairing their nets. In the village, we ate freshly caught fish grilled over a charcoal fire scraped out in the sand. They served us thimble sized cups of Macadam-like Arabic coffee and we distributed Marlboro cigarettes. The children gathered around us, the flies around them, and in thirteen months we never saw one village woman.

It was not all fun and games. We were part of a large contingent of British and American military forces - army, navy and air force - stationed around the Persian Gulf, or as we call it today, the Arabian Gulf.

Our respective governments had not sent us there for a holiday.

We were there for a reason.

Chapter 7

Sun, sand and oil

There are three things to be found in abundance in Arabia: sun, sand and oil.

The British had been present in the Persian Gulf since 1853. The area we were to operate in was referred to as the Princely States. These princedoms had been placed under the jurisdiction of Colonial India, an annex of the British Empire. Historically, Britannia had held sway over the world's largest empire, one on which the sun never set. Up to that time, most of the coastal Arab sheikhs had practiced goat herding and piracy, the latter the British had quashed. Victorian and Edwardian Britain believed that God had placed Englishmen on the Earth to rule the less worthy members of the human race. Our leaders, 100 years later, were of much the same opinion. Today, the Princely States are known as the United Arab Emirates.

We were there to ensure that the West had access not to sun and sand, but to oil. Plus, something not often acknowledged, British and American forces were in the Arabian Gulf to ensure that the Soviets did not gain deep water ports and access to oil. A Russian naval presence of carriers, destroyers and submarines in the Arabian Gulf or in the Indian Ocean was what Whitehall, the White House and New Delhi all feared (the Pakistani's, true to form, were playing both sides of the fence). All Arabian sheikhs feared the doctrine of Karl Marx, and some were fighting local Soviet-backed insurgencies.

Our being there was not motivated by a need to defend democracy, nor had it anything to do with humanitarian principles. It was driven by the same interests behind the more recent 'wars of liberation' in Kuwait, Iraq, Afghanistan and Libya. It was, and is, about control, power, access to oil and the flow of capital. In 1960 the Middle East was in turmoil, and British taxpayers would have to foot the bill of military intervention for yet another silent war.

Those who died, on both sides, died for one deity: **God Oil.**

The House of Saud - They are probably the richest and most evil family in the world. Surpassing their partners in crime; the Rothschild's, the Vatican, the Mafia, the Bush Family, Wall Street, the World Bank and the United Nations.

The Saud were originally one of a number of nomadic desert tribes, who had waged sporadic guerrilla raids against Turkish occupation. When Arabia was freed from the suzerainty of the Ottoman Empire, after World War 1, the Saud sent small patrols of camel riding henchmen across vast areas of the peninsula. On entering villages, often their earlier allies, in the depth of night, they dragged the headman out of his hut and cut his throat. They then dumped his body in the middle of the village square, to the sound of gurgling jugular blood pumping onto the sand. Widows howled, dogs barked and the village was left in shock - and firmly under Saud rule.

This tribe of illiterate goat thieves ended up controlling a massive area of waste sand that no one really wanted. Even so, the British were not particularly happy with developments, as the House of Saud could easily gain control over the Trucial Oman Sheikhdoms. With this, they could control oil transport in and out of the Gulf. Whitehall wanted to avoid a war that would tie them down for years. So the powers-that-be arranged a meeting in London between the Saud *'royal-family'* and King George. They were accommodated in 5-star hotels, wined and dined at 'special gentleman's clubs', with access to unlimited malt whisky and escorted by skimpily dressed young ladies.

This was followed by an impressive display of British military power - land, sea and air. All of which persuaded the Princes of The House of Saud to cease their expansionism across Arabia. Giant hangovers and a wish to return home, to escape the cold of the English summer, led to their signing a series of treaty agreements.

Then in 1938, with the discovery of oil, Saudi Arabia went from being one of the poorest countries in the world to one of the richest. An irritating turn of events, yet at that time, America had enough homeland oil and Britain had control over the Trucial Oman States. It was felt to be more prudent to negotiate than to annex a vast sand wasteland by military intervention.

Today, from a financial and infrastructural point of view, Saudi Arabia is a super-modern state. The country is kept functioning by near slave like labour from Pakistan, Nepal, Malaysia and other countries who work in the most primitive of conditions, for symbolic wages with few, if any, human or legal rights.

The House of Saud has and still does finance the expansion of Muslim terrorism worldwide. They, in cooperation with the US oil barons and the US weapons industry, part-financed George Bush Senior's and Junior's presidential campaigns. Explaining why, just hours after the Twin Towers attack, a number of planes filled with Saud family members flew home unhindered, despite the total flight embargo over American territory. The film *Fahrenheit 9/11*, by Michael Moore, should be compulsory viewing for everyone.

The West's dependency on oil, coupled with Israel's and the Saud family's joint fear of Iran's nuclear ambitions, guarantees the continuation of this perverse alliance.

The Trucial Oman States - They originally consisted of eleven independent Arabic Sheikhdoms. Today there are seven states in the union, all led by a sovereign Sheikh. Each having their own Disneyland palace(s) and private army, their own bank account in Switzerland and their own very exclusive harem. The social standing of the Arab male is based on lineage, age and wealth. Wealth can be viewed by the number of oil wells, camels, sheep, goats, Porsches or wives he owns. In addition, the more male children sired enhances his status.

If a poor Arab goat herder sires a girl and cannot afford another mouth to feed, then at her birth she is simply taken from her mother and put out in the desert at night. Her crying, or smell of fresh placenta, will soon attract a desert fox or a jackal. Her fate sealed as she is dragged off across the dark sands to be eaten alive by a pack of hungry cubs. On the other hand, the Sheikh or rich Arab, has his wives or slaves to care for any excess girls.

This is not in the past tense. This is how the Arab male views himself - and the role of his woman. It is the same, be he residing in a palace in Arabia or sitting unemployed in a high-rise block in the West, praying five times a day and living off social welfare.

Once the men have completed their pilgrimage, Al Hajj, to Mecca, they are blessed by the prophet Mohammed, and become as one with the almighty: Allah.

The boys they sire reflect their masculinity. Yet, there are potential downsides for a sheikh, because in the harem are several mothers (wives and concubines) with many children. Here each mother, with nothing to do, builds up her elder son, with an eye on the succession. On reaching manhood, the spoilt and pampered young men actually believe they have some form of talent. While in reality they are lazy, arrogant and not particularly intelligent. The latter being a result of countless generations of inbreeding.

They also know that if they are to get a harem of any worth, then Dad must be removed.

Shoot a Sheikh - From 1950 to about 1970, a number of the friendly Sheikhs were experiencing situations of unrest within their own families, as well as from the various tribes within their domains. Shoot a Sheikh became a popular pastime. That nomadic tribes had no concept of borders nor any respect for the property of others added to the confusion. One Sheikh might covet another's oasis or try to block a traditional trade route, resulting in shots being fired. Somewhat akin to the family intrigue we see in today's TV soap operas, except with guns.

Yemen was in uproar and British forces had been expelled from Aden, tails between their legs, with death and destruction on both sides. Plus there was Dhofar, where warring tribes wanted independence from Oman. Add to this the unrest between Turkey and Greece, plus that the British were forced out of Cyprus, resulting in the island becoming ethnically divided.

The map of the Middle East was being redrawn as we watched.

Britain and America's dependency on oil and the need to deny Russia access to the same was solved by offering Saudi Arabia, Bahrain and the Trucial Oman States a treaty of protection. There was a need for a gradual 'face-saving' transition to self-defence. To facilitate this, and to pretend that the British and Americans were moving out of the Gulf States, a private security company was established. British and American soldiers, with experience in the desert, would suddenly resign from the military and step straight into very lucrative civilian jobs as security consultants.

Later, we would see the same model in Iraq and in Afghanistan. Multibillion dollar 'private' security firms were and still are waging war as the representatives of 'the free democratic world'.

Who was defending who - and why? - An Arab pilot was never allowed to fly a fighter jet solo. There was always a western jockey on board to ensure the son of the sheikh did not get lost, crash the plane, or bomb his dad's palace.

The West sold them the toys of destruction, so we could finance the development of our own weapon industry. Yet, the West never relinquished control. Should a Sheikh fall out with America, or their weapons be turned against western allies, say Israel, then their squadrons would be destroyed by the pressing of a button in a bunker some place in the U.S.A..

Inconvenient minor civil wars were escalating and were affecting oil security and as a result oil price stability. From the island of Bahrain, American and British forces would be flown into various troubled areas to support our favoured Sheikh against an uprising.

For example, in Dhofar, some tribes were seeking independence from a brutal, but western friendly despot. They were quickly labelled 'Communist insurgents', a 'truth' accepted and propagated by the non-critical media. They, the Russians, were seeking access to 'Indian Ocean' ports, a totally unacceptable situation for the ruling Arabs and the western powers. So, nationalists or freedom fighters had to be crushed at any cost.

Another area, and not the worst, was Muscat, where I suddenly found myself on desert patrol.

On patrol in Muscat and Oman, confiscating rebel weapons. (1968/69)

الطريق على ايفرست

Chapter 8

Midwife in the sands

In order to retain stability in the Trucial Oman States, small patrols, using three or four Land Rovers, were sent out into the desert. We were totally self-sufficient with water, food, fuel and weapons, enabling us to operate for weeks on end.

Our function was to establish and maintain radio masts, and to provide tactical support to the local authorities, by backing the Trucial Oman Scouts. The Scouts were an elite group of soldiers, who at one time numbered over 30,000 men. As the number suggests, they were not kept for ceremonial purposes. In our theatre of activity, they functioned as the police of the desert, and were usually led by ex-British officers. The soldiers were often Jordanian, a proud and efficient desert fighting people. The Trucial Oman Scouts, fondly referred to as the 'Desert Gurkhas', acted as arbitrator between conflicting groups, and if necessary, they would crush an uprising with great speed and unflinching brutality.

Often it was just Arab killing Arab, for seemingly no reason. One time about twenty people were killed over a disputed water hole. It was amazing how many people could be killed over the ownership of a scraggly goat. The wrong word by the wrong person could lead to a blood feud. The code of honour ignited by a lustful eye directed at a bare ankle could result in generations of killings, which continues until long after they have forgotten why they were killing each other.

007 of the desert - Our field leader was Sergeant Pibworth, affectionately known as Pib, a true legend in desert rescue, survival and warfare. Having spent most of his military career in the desert, he spoke Arabic fluently, as well as a dozen dialects. He held the respect of all who crossed his path, yet he never rose above the rank of Sergeant. He was no James Bond, but he partook and backed a number of hush-hush operations all across the Gulf.

A child is born - One evening two gun-bearing men approached our encampment seeking help for a woman who was having problems giving birth. Pib pointed at me, the youngest in the group, *'Come on youth, time to teach you the facts of life.'* The others smiled, as Pib rarely used such long sentences. I followed him like an excited puppy, and half an hour later we reached the village. Western men were not allowed to see the village women, so Pib simply put his hands through the dirty curtain that separated him from the mother to be. Here, with smoky oil lamps and a few towels, we started what was to be a two hour long labour.

He checked the baby's position, uttering reassuring words. I mentioned that the colour of the water I was heating left a lot to be desired. Pib snorted, *'If we don't kill her, the next child or a scorpion will.'* His second sentence, we seemed to be getting on well.

She was just a moving shadow on the other side of the curtain, the whole thing seemed surrealistic. There were the usual moans and screams, all by her, while Pib's Arabic grunts and English swear words did seem to help. Then suddenly, it was over, a child was born.

It was a boy.

Relief, a sigh, a fleeting smile, and another grunt from Pib as I passed him a wet towel to wipe the child after he had cut and tied the umbilical cord. He then passed the baby boy to the father, but as we were cleaning up, a sudden gasp stopped us. Pib sensed the mother was bleeding. He quickly stepped through the curtain, only to be savagely grabbed and thrown onto the floor. He struggled, but was held down. I straightened, not realising what was happening, only to find myself stopped by a curved-bladed knife. Pib had two rifles pointed at him. Body and soul stopped, we stood motionless, breathing heavily, arms by our sides. Guns were pointed at Pib and a knife at my throat. We watched as her ankles twitched outside her gown. We felt her anguish, listened to her sad-helpless sobbing as she, the men and we all realised that her life was gently ebbing out onto the cold mud floor as we stood there. Then there was silence.

The father had a son in his arms, which he raised towards Qibla (Mecca) and then, not looking at his dead wife or us, he turned and walked away.

We were led away and at the edge of the village their guns were lowered. There were uncomfortable smiles, a boy had been born, the village was blessed and they forgave the 'misunderstanding' and waved us goodnight. We walked back to our Land Rover camp, to our disturbed dinner of tinned Mulligatawny soup and jam-plum pudding.

'Damn their black heathen souls,' said Pib. His third sentence.

Then, the next morning, as the sun warmed the valley, we watched from the neighbouring hillock as two men scraped a shallow pit in the hard earth. Desert burials are quick and simple affairs. Almost before they were finished, a small procession of men carried her thin body, wrapped in white cloth and placed her in the pit, her face towards Qibla. Prayers were recited and each man in turn scattered handfuls of earth over her body. Then they covered it with coarse gravel and flat stones to stop the jackals from digging her up. A stone, with no inscription, was erected to mark her last resting place.

Far to one side stood a woman, maybe her mother, dressed in an all covering black abaya (full robe) and three girls, probably the dead woman's daughters, all silently motionless.

Undoubtedly Pib took with him a number of messy secrets to the grave. Thirty-three years later I noted his five line obituary in Time Magazine and thought, *'I bet Whitehall is glad Pib is dead and buried.'*

Next day, as a token of gratitude for our having assisted with the successful birth of a boy, we received a basket of plump prime dates, a bowl of yogurt and a playful kid goat. The later we grilled for dinner that evening.

The Khanjar *is a curved Omani knife sharpened on both sides.*
It is taboo to take it 'fully' out of its sheath, unless you intend to use it.

Chapter 9

Days on the Rock

After thirteen months in the desert, I returned to Britain. Again lady luck smiled as I was posted to R.A.F. Valley in North Wales, near my home town, Holyhead. Valley was a training camp for fighter pilots. In addition, it was a top secret Nuclear Aircraft Diversion facility. Here, in the event of a war readiness *Code Red* being issued, a nuclear bomb carrying Vulcan aircraft would be diverted, on 'war-standby'.

I worked nightshifts, which gave me three days off, allowing a lot of free time to climb.

Solo sea cliff climbing - Unfortunately, all my friends had moved away to college, or their interest lay in mountain walking, resulting in my having problems finding people to climb with. So I devised a crude system of slings and bits of rope to protect myself when climbing alone. It was a bit dicey: *'Maybe it will work, maybe it won't.'*

This system allowed great freedom to climb alone on what to me were difficult routes like *Bezel, Central Park, Fifth Avenue, Times Square* and *Diogenes*. Often I chose not to use any slings or ropes, experiencing eerie climbing on classic routes like *Rap, Pel, Bloody Chimney* and *Rift, Puffin* and *Pantin, Simulator* and *Imitator*.

Descending the steep grass slopes to reach the base of the giant lunar shaped wall situated opposite the South Stack Lighthouse, with intent to climb *Green Slab* in the Mousetrap Zawn at night. Under a full moon, helped by the ghostly lighthouse beams, totally on my own, was my walk on the Moon. Neil Armstrong had millions watching him, I had 1000 irate gulls and guillemots for company.

Ed Drummond and Dave Pearce on the 1st ascent of Dream of White Horses, 1968. Paul and I did the 5th ascent the next year. Photo: Leo Dickinson.

Some may feel it is irresponsible to climb alone at night, but I was young, enthusiastic and indestructible.

Plus I had a Batman cape that allowed me to climb one of the finest and most spectacular climbs in Britain: *Mousetrap*, again on my own, no partner, no ropes. Today, I shudder remembering tentatively stepping past a nesting bird so that her one egg would not tumble into space. Should I fall, my body would spin into space and hit the rock at least twice before landing in the sea and certain death.

In truth, I had never questioned the *'why'* of climbing alone, but here, on reaching the last hard move, 200m above the sea, with only space below my feet, for a half second I pondered the *'why'*. Probably, not questioning the *'if I fall and die...'*, but the more a fear of embarrassment should I need to be rescued. How's that for logic?

Doubting or questioning was probably more dangerous than steep loose rock - for it took away the flow, and blew a hole in my Batman cape. It was safer that the senses were addicted to the touch of raw rock and a state of *no doubt* prevailed.

Another great day was doing the fifth ascent of *The Dream of White Horses* with Father Paul Sadoli, a Catholic Priest from Liverpool. The Dream had first been climbed by Ed Drummond and Dave Pearce (1968), and Father Paul had secured a rough diagram from Ed with scribbled notes. *'Two days... sky hook, loose rock, overhanging, friable...!'* We started this amazing line, with upward and sideways progression being helped by Paul reciting mantras in Latin.

No one knew we were there, so if stuck we would hang until we died of starvation. It was handy having a priest along, as he could administer extremunction, the last death rites - before our eyes were pecked out by hungry seagulls!

Llawder - Mike Shannon, the officer in command of Air, Sea and Mountain Rescue at R.A.F. Valley, and I used to climb together. Then one day in 1969, he showed me his secret sea cliff: Rhoscolyn!

This rich red cliff had lain totally hidden from the climbing fraternity, with Mike having spotted it from a rescue helicopter. As the climbing was far too difficult for him, he gave it away! What a generous present. I quickly introduced others to the crag, and together we climbed about thirty new climbs: *Icarus* with Pete Buxton (Joe Brown, the greatest rock climber in British history, found the crag by accident the next year, and did the third ascent of Icarus. He respected our being the 'first to find it' by not telling anyone. He could have easily done a dozen new routes, but he left the crag for us 'local lads' to develop - a true gentleman).

Icarus was followed by *Truant* with Dave Birch, *The Viper* (now called *Centerfold*), *E.P. Special*, *Cocaine* and *El Dorado* with Lou Costello, *Wild Rover* on my own - all the latter being graded Extreme. *Symphony Crack*.... etc. etc. Each name tells a story about the climb, or about those who first climbed it.

I baptised the crag *Llawder*, which sounds very Welsh, but is simply 'red wall' spelled backwards. Then in 1970 I wrote a small guidebook to the climbs at Rhoscolyn and to those on Holyhead Mountain.

Then I left for Norway in 1971.

Me making the first ascent of Truant, grade V/VS, 1970, Rhoscolyn. Photo: P. Sims.

NORWAY
Land of the Midnight Sun

Stetind. Winter. Photo: Torgeir Kjus.

Stetind - Norway's National Mountain, 1392m, may not be a Himalayan giant, but it raises itself impressively out of the fjord.
First climbed by F. Schjelderup, C.W. Rubenson and A. B. Bryn in 1910.
Ste means anvil, and **tind** means peak - the **Anvilpeak**.

W. C. Slingsby, The Father of Norwegian Mountaineering, said it was the ugliest mountain he had ever seen! I disagree.

Climbing Stetind with Anne Grimstad. Photos: Johnny Haglund.

Chapter 10

Cosmic Atomal Top Secret

Back at work my speed and accuracy in writing telexes was way above average, a skill that interested the powers that be in the R.A.F.. To this end, the security services instigated a background check into my character. They found no criminal records, financial problems nor suspect sexual tendencies, so my security clearance was upgraded to Top Secret. This led to a posting to the Allied Forces Northern Europe headquarters just outside Oslo, Norway.

Most people know that polar bears walk the streets of Oslo, that Norwegian babies are born with skis on their feet and that the sun shines all night in the summer. Only the last one rings true, but what attracted me was the Troll Wall. Europe's highest and steepest mountain rock wall - one kilometre in height. Unfortunately I was not there to climb, but to work on a Top Secret Nuclear Deterrent Programme.

The main NATO communication centre was situated in a gigantic cave system deep inside Kolsås Mountain, just outside Oslo. As an ordinary telex operator, I was on the very bottom rung, but I did have the interesting job of typing the weekly Top Secret Situation Reports. Our office was the Operations Centre, while the War Room was placed deep in an area known as The Bunker. Here, there was a giant caravan resting on thick steel spring coils. It was fitted with state of the art communication equipment, sleeping quarters, toilets and showers, along with its own kitchen and an independent air, water and energy supply. All to ensure that, if the Russians dropped a nuclear bomb, we would survive five to ten days to continue the joint annihilation of our respective people. Meanwhile, our families would burn to crispy toast along with everyone else.

We played our war games and they played theirs. Both with paper ships and paper planes and we followed what the Russians were doing in the far north. There were daily reports about their military aircraft, codenamed Bears, flying over the Barents Sea. Our fighters intercepted them, the pilots waved to each other, and then they both flew home for lunch. Our intelligence reports showed that the Soviet

Northern Fleet consisted of rusting U-Boats and decrepit warships that would probably do more damage to themselves than to us. Plus, we had our 'tourists' who reported military movement of Russian and Eastern Block forces, all the way from Poland to Siberia.

NATO's Giant Swans - Nuclear bombs were permanently placed in a fleet of Vulcan aircraft, based at R.A.F. Brize Norton in England. On the ground, the Vulcan looked like a cumbersome pregnant Albatross, but once airborne it was like a giant swan. In the event of a flash code RED, the Swans would scramble from Brize Norton and disperse to different airfields around Great Britain. At their new war alert stations, all crews waited, hopefully to be recalled by code STANDOWN.

Or, if they received code SCRAMBLE, twenty aircraft would take off to predetermined positions in the sky. Again, with the prayer for the recall back to base and normality.

Alternatively, high in the sky, code STRIKE could be received, and things become serious. A double sequence code would be checked manually between the pilot and co-pilot, an eight second process. If the codes did not match, then it was an exercise, and they would turn their aircraft and return home. If the sequences did match, then we were at war! They would turn silently in their seats to simultaneously press two red buttons on the console.*

This heralded the start of the end of civilisation. With their loads of annihilation pre-programmed for pre-determined targets, the Swans winged their way towards Russia and Eastern Europe.

There was no turning back. Their targets were military installations or high-density population cities. Once over the target area, the bombs would be automatically released. Millions would be deep fried. Grandparents, mothers and fathers, brothers and sisters, young and old, the good, the bad and the ugly. The nuclear bomb treats everyone equally.

Toys for the boys! - Polaris was a two-stage solid fuel nuclear armed missile on which America had spent billions of US dollars to develop. As the programme neared completion, their Defence Department realised that the missile's range was too short to reach enemy targets! If they were to penetrate the ABM-defence screen around Moscow, they had to get closer.

---

*That the navigator had a loaded pistol to ensure the two pilots did press the buttons, I cannot verify.

This was solved by placing Polaris aboard four British ballistic submarines, three of which were always on patrol armed with sixteen nuclear missiles each.

Vulcan aircraft *Polaris*

Let us not forget another nuclear delivery system available to the western arsenal, but not controlled by NATO - the independent fifth nuclear country, France. The French nuclear arsenal was, and is, frighteningly powerful and still more secret than secret.

Mary Poppins - The American umbrella of radar listening stations in Iceland, Greenland and Alaska was backed by an impressive arsenal of long-range missiles with nuclear warheads. The US-based missiles were aimed over the North Pole towards Russia, while the British-based missiles were aimed at the Eastern-Bloc countries.

'Spy City' - situated up on the Greenland icecap. Photo: Christian Eide.

Friendly Bombs - In Europe, Neutron Bombs were to be placed on moveable rocket launchers by 1970. These were officially categorised as friendly hydrogen bombs or Enhanced Radiation Weapons. Which in layman terms is a fission fusion thermonuclear bomb designed not to damage property, but only to kill people via non-destructive neutron radiation. This controversial program had been stopped by Congress, and re-introduced secretly by President Ronald Ray-gun. A comforting thought, that if one is to be eradicated then it is better done by a friendly bomb!

But back to our Flying Swans; I ask:

'Once the bombs were dropped, what then?'

What were these young pilots and their crews supposed to do? They were the sons and daughters of ordinary people, with families of their own, their children attending kindergarten and school. Were they to return to the green fields of England or Connecticut and live happily ever after? Was Ivan to return to the rolling wheat fields of the Crimea after he had dropped his nuclear eggs? At the end of the day there would be no place to call home, there are no winners!

The Russian Northern Fleet - The Soviets and their allies, with recent memories of the horrors of the Second World War, believed that NATO planned to attack and annex their territories. Western European political and military minds believed that the Soviets would enter Europe through Poland using conventional land forces, to spread the doctrine of communism. To meet this threat, NATO would defend Europe with soldiers, tanks and planes.

From the Baltic Sea, Russian battleships would attack ports in Denmark and Holland. At the same time, the Russian Northern Fleet would steam down from Murmansk and Arkangle by-passing heavily defended Britain and France, to attack the weak under belly of NATO, Spain and Portugal, thus creating a third battle front. In the Mediterranean, Russian nuclear submarines would be standing by, to neutralise Italy and Greece.

Northern Shield - It was general knowledge (at least to the sleep over girlfriends of most NATO personnel) that Northern Shield was the codename for the dropping of nuclear bombs on the Russian fleet off the coast of northern Norway.

The government of Norway did not allow NATO to base nuclear fission material on its sovereign territory. So if NATO deemed it necessary to drop The Bomb on or near Norwegian territorial waters, the Norwegian Prime Minister and his cabinet had to resign and transfer executive power to King Olav. He is Commander in Chief of Norway's armed forces, with the codename Queen Bee.

Queen Bee would then transfer military control to the NATO commander at Supreme Headquarters Allied Powers Europe, in Mons, Belgium.

Then, it would take no more than fifteen minutes before the total destruction of the Russian Northern Fleet by British Swans and Polaris submarines was accomplished.

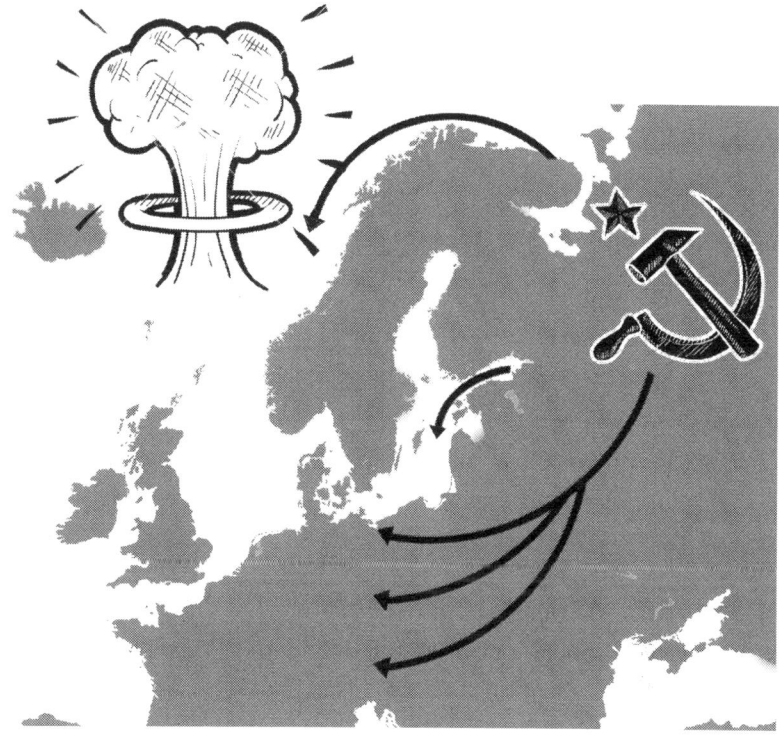

The bombing would result in a catastrophic wave of tsunami proportions, leaving thousands of dead in Island and all along the north Norwegian coast. Nuclear fallout would be spread by winds and distributed via the Gulf Stream throughout the Atlantic Ocean to enter the human food chain.

The mushroom cloud would be visible as far as London and nuclear pre-cooked fish would rain down on the streets of Europe.

Just add chips, salt and vinegar.

Chapter 11

The Normal World

When we were not planning the destruction of the world, I lived with my Norwegian girlfriend, Sissel. We shared a flat in downtown Oslo with U.S. Air Force Sergeant Mike Cosby. Sissel and I were both engaged to other people when we first met. On meeting, we totally forgot everything else. Our first hurdle was to get un-engaged as fast as possible.

Sissel studied to be a dental hygienist and I worked for NATO, which really meant I hardly worked. We military personnel were young, full of hormones, overpaid and had access to seemingly unlimited duty free booze and cigarettes. When not working or partying, I went climbing. Luckily, my Commander allowed me time off, as a military exercise, and I ended up working at the Norwegian Mountaineering Centre at Turtagrø Hotel as a climbing instructor.

Mike had served two active tours of duty in Vietnam and had twelve bits of shrapnel in his leg. In addition he was generally disillusioned by both God and mankind. As part of the GI-Bill, an American law providing higher education to serving soldiers, Mike was studying psychology part-time. This and other courses were organised by the University of Maryland's overseas campus, and were open to all American soldiers. I applied to join Mike's class and was accepted. The visiting lecturer, Professor David Heart, soon moved in with us, and a boozy 'Mad Poet's Society' was formed. All subjects and theories were discussed, dissected and challenged.

My other passion was reading the works of the likes of Orwell, Huxley, Defoe, Anatole France, Shakespeare and those who exposed the political and social structures of their times. The world was becoming clearer: I was stuck in the middle of a mire of death and destruction, lies and deceit. I began to question the vast waste, incompetence, inefficiency, irrelevancy and corruption in the military machine and in the shallow political systems that the military ensured the continuance of.

There is no such thing as democracy - I wanted out and, as usual, Mike had the answer, *'Inform them that it was your sister who signed the papers. Therefore, the contract is legally invalid and you are a political prisoner!'*

I vetoed this, thinking it might get my sister in trouble.

His next plan was that I should lose my Cosmic Atomal Top Secret security clearance. This would happen if one had economic difficulties, problems in one's marriage, drink or drug dependency traits, was homosexual or if one was a pervert. All of which might allow Soviet secret agents to turn you into a spy. Inconveniently, I was none of these, so we conceived a new master plan over bottles of beer and buckets of red wine. It was based on the theory that my strong dissatisfaction with the military would cast serious doubt as to my loyalty to Queen and Country. This, as the theory went, would see me losing my security clearance resulting in my being discharged.

I remember standing in front of my commanding officer and it went something like, *'I am unable to neither be a part of or accept the inefficiency, irrelevancy nor justify the vast costs related to the military when people are dying of hunger in Africa. I must remove myself from a system that supports corrupt and self-serving so called democratic political regimes, where in fact they oppress their own peoples.'*

I was immediately neutered and denied access to security sensitive workplaces. Did they doubt my loyalty, or did they think I was a raving lunatic? Irrelevant, because two weeks later I was returned to England where I received an honourable discharge. Free!

Sadly I lost contact with Mike, but ten years later learnt he had qualified as a psychologist, and was counselling inmates on Death Row.

Out of the R.A.F. and suddenly unemployed, I returned home to Holyhead and Sissel followed a few weeks later. She worked in my mother's restaurant as a waitress and I climbed during the day and worked evenings in my brother Philip's pub. After five months of climbing and living on a diet of greasy fish and chips, washed down by brown ale against a background of depressing Welsh weather, we decided to return to Norway.

The move back was in part driven by a spreading depression that was hitting Britain as unemployment increased, immigration was getting out of control, and the trade unions were strangling British manufacturing industry.

Maggie Thatcher had not made her mark - yet!

Lost in Translation - Back in Oslo we were broke. Sissel found a job as a dental assistant. I, on the other hand, could not speak Norwegian, had no civilian job skills, little relevant education and no legal work permit. Add to this that we were not legally married, lived in Sissel's mum's attic and used a red plastic bucket as our toilet, then the future looked pretty bleak.

The first priority was to apply for a work permit. For this one needed a job contract with a registered company. Per Gaarder, the father to my climbing partner Petter, offered to employ me as his company driver. This was both noble and brave, as they did not need a driver, nor did I have a driver's license.

To solve the last problem Petter simply borrowed his father's car.

Lesson one: Ask permission first.

Lesson two: Learn Norwegian!

We were driving towards a bend when Petter said, *'Litt sakte her.'* I kept pumping gas. He repeated, *'Sakte her.'* I looked at him and smiled, thinking he was praising my driving skills. His face went white as he realised I had not understood that *'sakte'* meant *'go slow'* - then everything went very fast.

'SLOW DOWN, FOR CHRIST'S SAKE!!!!!!!!!!!!'

Petter grabbed the handbrake and we took off. As you guessed, handbrakes don't work in flight and his father's SAAB flew! Silence, unreal flight through light, we landed surprisingly gently in a field, much to the amazement of a herd of cows.

The job offer was rescinded.

A few weeks later I joined the great proletarian hordes working on a production line, manufacturing hi-fi radios and cassette players at Radionette. The next move was to Stolt-Nielsen Shipping as a telex operator.

In 1971 Petter (sitting) and I worked part-time as climbing instructors at The School of Mountaineering, Turtagro Hotel, in the Jotunheimen. The hotel became my second home in Norway for the next 30 years.

Mountain Magazine - Ken Wilson, the editor, asked me to try to sell the magazine. First stop was the main climbing shop in Oslo, SportCo, to meet a guy called Noble. He asked my opinion about his shop's climbing equipment.

'Load of rubbish; your ice axes are too long, safety gear is ok for limestone, which is non-existent in Norway. Your ropes need to be a thinner diameter and they should be 20m longer.'

Instead of throwing me out he invited me to a major outdoor equipment exhibition in Germany the next week. In a whirlwind four days we established a number of new agency agreements - Berghaus, Edelrid, Troll, Clog, Salewa, Scarpa...

Soon I was ordering equipment, followed by shop staff training and designing advertisements. The next three years saw me working late night telex shifts, to make a living, and flying off to sport exhibitions and factories around the world: Harrogate, Milan, Cologne, Munich, Chicago, Salt Lake City, Hong Kong and Tokyo.

By this time Sissel and I had moved out of her mother's attic into a flat of our own, with a real toilet. We had work, had got married, and then along came Therese - a healthy and happy baby girl.

Unfortunately, the hobby at the shop was time consuming and led to criticism from my real employers. In addition, it put a strain on our marriage. So I decided to quit the sport business and informed Noble.

To which he replied, *'Leave the telex job and run the shop full time.'*

Remembering the little old ladies and their seaweed cakes, I turned him down flat. He countered, *'You want freedom, but you need security. Here you can have three months off with full pay.'*

A deal was made on a handshake, three months and three weeks free - a full year's wage, and a free hand to run the mountaineering department. This was a gentleman's agreement which functioned problem-free for seven years. I had itchy feet and wanted to travel - now Noble had given me the chance.

THE SILK ROAD

A view on the Hagia Sofia, Istanbul.

A boyhood dream

To journey from London, through Asia, to reach the fabled city of Kathmandu.
Not in the footsteps of the great explorers and conquerors, but as an ordinary bus tourist.
Should you have a dream, but have no plan, then do a Nike:
Just do it.

Surfi whirling, Istanbul.

Houseboat on Dal Lake, Srinagar.

The C. of E. Bishop of Pakistan.

Taj Mahal, Agra.

The night's catch.

Village allure, Goa.

Chapter 12

The last of the pretend hippies

The books *Annapurna* and *Seven Years in Tibet* were always in the back of my mind and new ones entered. The travels of Marco Polo and Alexander the Great, along with the exploits of Stanley and Thesiger, all inspired. In 1977, I left the shipping company with intent to start work in the sports shop two months later. Sissel had agreed to the master plan: I travel overland to India and get 'it' out of my system. Then on returning, I would settle down to a normal family life.

A group of twenty or so loosely connected people had bought a second hand 46-seater bus in London and I joined them. They had ripped out the seats and refurbished the interior, adding sofas and tables and had installed a music system bought from a bankrupt strip club in Soho. With our own disco and smelling of paint we crossed the English Channel in a storm, picking up 50 cases of duty free Heineken on the way. Then we drove southwards through rain-swept Europe to bathe on the sun drenched beaches of northern Greece, before travelling east to enter Turkey.

Istanbul was the mystic Constantinople of my books, capital of the eastern Roman Empire and gateway to the Silk Road. Here we were bedazzled by the impressive Hagia Sofia Mosque. The sights, smells and sounds of the bazaar were exotic and fascinating. Almost alien, this was the crossing point of a million stories, where west meets east. Darius, Alexander, Roman Emperors, Sultans and Pashas, the incredible scholar and much travelled Ibn Battuta and, of course, our own Agatha Christie.

Crossing the Bosphorus, we entered Asia Minor, to be greeted by the swirl of Dervish Dancers. In side street cafes we listened to wrinkle-faced Kurdish men telling stories of a nation unjustly and savagely dispersed across five countries with no home to call their own. We learnt of the hideous slaughter of the Armenians, a state sponsored genocide by Turkish forces. A hideous crime, which today's Turkish state still denies, despite irrevocable evidence. We listened, nodded our heads and believed we had something to contribute.

Playboy visa - To smooth the border crossing into Iran Fred, the bus driver, discreetly slipped a brown envelope with four copies of old Playboy magazines to a guard. The guard half pulled them out, looked up, a few tense moments, then he nodded. Fred had a way of expressing authority and comradery at the same time.

In another room, we later heard, girls on another bus were half stripped, by sticky fingered custom officials - their bosses probably jerking off in the next room.

Iran was more than just arid desert, there were great mountains in the distance. We viewed temples and the remains of ancient empires and swam in the world's largest lake, the Caspian Sea. It was so large and salty that the ancients had regarded it as an ocean. Now families played in the waters and we watched as Iranian women paddled ankle-deep, decked head to toe in their black chadors and, from a distance, they resembled penguins walking on water.

In Tehran one of our girls was punched, kicked and spat upon, seemingly because she wore short sleeves and had long peroxide-blonde hair - a western whore! We were naive in believing that western norms prevailed because the Shah was progressive. For when we stepped into the backstreets, we were confronted by totally different values. The police were polite, but seemingly they held the attitude that, 'She got what she deserved'. To them the Shah was a puppet of the West, and western women were whores.

We were the last of the pretend hippies crossing Asia. Selfishly and naively ignorant of the cultures, peoples and lands we passed through. A lesson had been learnt, but was soon forgotten.

A few days later, we were eating melons and dates under palm trees, listening to the music of Jimmy Hendrix. Sometimes we stayed in small villages, sometimes camping, grilling our food and eating out under the stars. If the wind blew or a downpour hit us, we bought kebabs from a street vendor and ate in the bus. There were love affairs, break ups and emotional trauma, understanding and comradeship, laughter and tears, highs and lows. We travelled in a pink time-bubble through strange lands during equally strange times, genuinely unaware of just how irrelevant we were.

Occasionally, we stayed at a city campsite or guesthouse to gain access to a shower. One evening, half the group had gone to bed, when we heard the sound of fireworks, which we assumed was a part of a local festival. Wrong! In desperate haste, our host turned off the lights and herded us to our rooms. After the shooting there was total silence, to be broken by orders shouted, the slamming of doors and the revving of engines. Then again silence. A dog barked. Some of our girls cried and we men stood by, helpless and pretty bloody useless.

The next morning we learnt that the Shah's soldiers had opened fire on a group of students, who had gathered illegally.

We had a silent breakfast and continued our journey.

Afghanistan - The Last Frontier - We crossed the border into Afghanistan, at a time when the country was in the throes of yet another civil war. This was before the Russians invaded - where Afghani killed Afghani.

James Michener's book *Caravans* was my companion as the countryside flew by, the towns of Herat and Kandahar merging. This was Marco Polo country, the last frontier, barren, inhospitable and unforgiving. We flashed through in our bus, swilling our lukewarm Heinekens as we journeyed across a landscape that straddled east and west, north and south. This strategic position is one of the reasons Afghanistan has been fought over by Darius, Alexander, Genghis Khan, Kublai Khan, Queen Victoria, Leonid Brezhnev, George Bush and his lapdog Tony Blair.

In 1977 we were simple tourists, passing through before the Russians officially entered the country in 1979 to liberate the Afghans from themselves. For us everything was new, unique, exotic and exciting. We, the privileged white, only experienced the occasional inconvenience, like a blocked toilet. Others suffered pain and death simply because they lived in their own country.

The Russians in turn would be beaten by the harshness of the land and the tenacity of its people. The Afghans would then inherit Pax-George Bush - confusion, hate, destruction, desperation and death. Orchestrated by his Vice President and fronted by his Minister of Defence. Their political positions allowed them to be pimps for a multi-multibillion dollar 'war machine'. To receive their pre-agreed 'blood money' once they stepped out of office.

What have they done with their 40 pieces of silver? They were responsible for the butchering and the maiming of hundreds of thousands of innocents; their own and other nationalities. Murderers are put in prison, some are executed, yet Prime Ministers, Presidents and politicians who murder and pillage their own and other nations are rewarded with highly paid, high-profile positions, book contracts and pensions.

The Conspirational Theory is no theory - it is reality. The Industrial War Machine creates death and destruction, to be followed by the vast movement of refugees, creating a humanitarian buffer to provide a smoke screen, known as the Diplomatic Industry, which inturn prolongs the war, dividing the spoils, and in so doing, paves the way for the Foreign Aid and Development Industry.

Before one war is over a new one is being conceived and managed, its outcome pre-determined, and the three 'industries' continue to play their role in this never ending circle. All three having a common agenda: to siphon billions of dollars into political and corporate bank accounts.

As long as we in the west refuse to acknowledge the central and important position the caste and tribal system holds in eastern countries, nor understand the 'why' behind its evolution and the reason for its continuance, then we have a problem. For example, to expect the Afghans, in his own land, to change and adapt to our model and norms with the resources available and within the timeframes we impose, is totally incomprehensible - Mad Hatter territory.

As to be expected, NATO filled the gap, only to retreat with its tail tucked between its legs. While the arms and aid industries continue to earn their countless billions. The West is still committing future 'aid' to prestigious development projects for this same feudalistic, corrupt and inefficient Afghan government. Who, in all probability, will ally itself with Al-Qaeda to butter itself against the rising influence of the Islamic State - or, alternatively, will hop into bed with IS - creating a new caliph. All financed by western aid.

Explain the logic in that to the man in the street, whom through his taxes finances war after war!

Cock fighting in Kabul - Let us return to 1977 and Kabul. Here we stayed in a 5-star-minus guesthouse, ate simple meals and witnessed deadly cock fights in backstreet tea shops. One evening the curfew was enforced early, leaving two of us stuck in a tea shop. Under cover of darkness, gun-bearing men appeared, looked at us, the café owner nodded and we were ignored. They talked in low tones and then left as silently as they had arrived, carrying their rifles as naturally as a London businessman would carry an umbrella.

The next day, I booked myself into the Kabul Inter-Continental, for a shower, shave and shampoo. Clean sheets, with no bed bugs, silky soft toilet paper, air-conditioning and a few hours in the shallow end of the swimming pool were bliss indeed. Some hippie!

Please, keep this our secret.

We drove through the Salang Pass towards the fabled Khyber Pass with a clear message ringing in our ears, *'Do not step outside the bus, someone will shoot you.'*

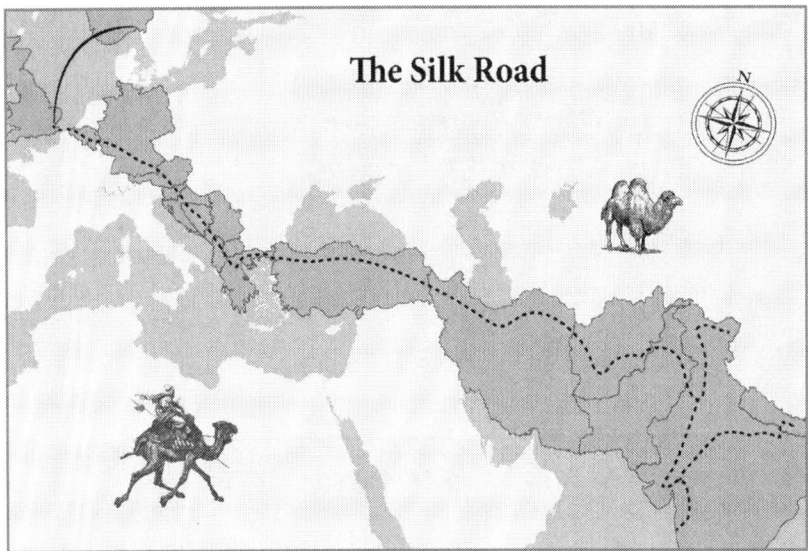

The name Khyber is derived from Hebrew, meaning fortress. Over hundreds of years the local Pashtu have levied a duty on both goods and people who pass through the area. This duty, along with the manufacture of guns and drug smuggling, are the main industries. Hence it is controlled by terror and only the Pashtu can reign. Virtually every man carried a gun that seemingly was manufactured dating back to the British Raj, or they were handmade over small charcoal furnaces in tiny workshops by the roadside. During the daylong crossing we did not see one woman - just the occasional shadow.

We travelled via Peshawar to Rawalpindi to reach the hustle and bustle of Lahore, the city of Kipling. From here the road led us into India, to reach the holy city of the Sikh's, Amritsar and the Golden Temple. The Sikh people embody the qualities of Sanat-Sipahie or saint-soldiers. Strong and proud, they strive to defend the oppressed, irrespective of their religion, colour, cast or creed. They are warriors of renown, yet there is a strange similarity with many facets of peaceful Buddhism. They practice a combination of humanity and humility, and have an immutable belief in the strength of moral spirituality.

Chapter 13

Playing golf with the Bishop of Pakistan

From Amritsar, the others continued through Northern India towards Nepal. I did not wish to cross India in the monsoon rains, so chose to spend three days at the Golden Temple. Here I took communion, slept on a reed mat and received simple food, in return for my scrubbing the marble floors. Then took a rickety bus north to Kashmir and the Garden of India, a luxurious haven where the Mogul Emperors had held their summer court.

At the bus stop in Srinagar, the district capital, hundreds of people were trying to rent out their houseboats to too few tourists. These flat bottomed boats, like floating caravans, lay moored on the beautiful Dal Lake. In the midst of the pandemonium, I chose a man who proclaimed, *'I pay taxi boat, you don't like my houseboat, I take to other until find one you very much happy with.'*

The sort of 'no win no pay' deal I love, and Bashir was true to his word. I stepped off his small water taxi and entered a large living room, where I sunk virtually up to my knees in the plush wall-to-wall red carpet. There was a ceramic charcoal burner, a chandelier and a piano, plus two luxurious bedrooms. One of these had a four-poster bed and an incredible king-size porcelain toilet with gold gilded handles. I had to stay!

The Bishop of Pakistan - The next day, in a backstreet chai-shop, I bumped into an elderly gentleman of obvious European descent where a nod turned into a conversation. He was English and, to my surprise, the Bishop of Pakistan. He semi-resided in India and, once he learnt I had just travelled overland from Europe, wanted updated information on Afghanistan. With this as his aim, he invited me to play golf the next day and enquired about my handicap.

'A compressed spine in three places, and a whiplash neck,' I replied.
To which he roared with laughter and I looked puzzled.

The opportunity to have 'Spy for the British Empire' on my CV was too good to miss. This man of God left nothing to divine intervention, as he employed two caddies. One to carry his golf clubs, the other to fetch the balls that had not landed where God intended and then place them in a more favourable position for the next hole. I, on the other hand, had to carry my own clubs and look for lost balls in the rough, expecting at any time to confront a man-eating tiger.

The Bishop held evening court for Srinagar Society on the flat roof of his houseboat. Here his turban-clad servant poured generous measures of Gordon's Gin, with a splash of tonic. The latter a time honoured antidote for malaria. In addition to small banter, he entertained us with real-life stories from the two world wars.

'Churchill, bloody bully, an uncouth man. Gave too much away to Roosevelt and Stalin, trying to be greater than Caesar. Let too many of our boys die... Damn him...'

Then he would go silent. Had he lost a son, or...? Who knows the pains and sorrows of those we meet along the road? Then he took a deep breath, straightened his back, slapped his knee and the conversation went on as before.

He attracted a strange and motley crew to his gin sessions. From retired colonial administrators to avant-garde intellectuals, there were European widows, gun runners and sari-clad ladies of ill repute. *'She had been the lover of the Raja of Puridistany and, before he was cold in his grave, she seduced the elder son... look at those diamond rings!'* Then, discreetly pointing at a wizened prune of a lady, who smelt of formaldehyde, he noted with respect, *'She, would you believe, bedded the Viceroy, well, so they say.'*

Ladakh - Little Tibet - On the Bishop's advice I decided to visit the neighbouring district of Ladakh, known as Little Tibet. He colourfully described it as being medieval, picturesque and, *'...more Tibetan than Tibet, since those bloody Chinese had not invaded it.'* It was also the starting point for some of Svein Hedin's exploratory expeditions into Central Asia - a walk in this great man's shadow was too much to resist - he was my Livingstone.

The bus journey necessitated a 250km road trip, which took two days, along the Big Dipper of roads that switch backed up and down the mountains. It was most disconcerting to spy a bus lying on its side 500m below us. One could imagine it bouncing down the mountainside, you could almost hear the passengers' screams.

Perhaps one should not complain, because before the road was built it had taken Hedin 16 days to walk from Srinagar in Kashmir to Leh, the capital of Ladakh. Once there, I was greeted by the dusty streets that a thousand other travellers and pilgrims had passed through. The city was more a town, a mishmash of buildings in the shadow of the king's palace. My eyes were treated to an array of Buddhist chortens, prayer flags and maroon dressed monks set against a backcloth of snow capped peaks.

Rabies - Officially Ledakh had only opened to western tourists the year before, so there were few other western faces in town and seemingly no lodges. I started looking for a place to sleep and ended up on the flat roof of a local café. After a typical Tibetan dinner of thukpa (noodle soup) and momos (meat and vegetable dumplings), I met up with other westerners and tried my first (and last) 'Kashmiri Shit' joint. While the others were giggling and being philosophic-silly, I sat stone-cold sober waiting for the drug to take effect. Sleep took over and I awoke some hours later, thirsty, groggy and desperately in need of a pee. On the way to the toilet, the guard dog bit my hand, but on realising I was not trying to break in, released its grip.

The rest of the night was spent hallucinating weird technicolour dreams as the Kashmiri Shit kicked in. I was flying with angels on fluffy clouds, to be suddenly chased by a pack of foaming at the mouth two headed dogs. I awoke sweating, vomiting waves of green slime, eyes coming out of my head and my mouth feeling like a baboon's bum.

Totally convinced I had rabies and was doomed to an agonising death.

My newfound French friend Patrick gravely summed up the situation. *'To be bitten by zee mad dog with zee babies is zee terrible ting, mon ami. If you go woof-woof I cut your throat and put you out of zee misery. There is no cure for zee dreaded babies'.*

I felt better the next day, having neither rabies nor the dreaded *babies.* We parted. Patrick disappeared to places unknown, while a series of buses and trains took me south to explore the impressive fortresses in the desert state of Rajasthan. Here I went camel riding, slept in the sand under an open sky and drank 'camel-piss'. The latter is a form of rough-poteen the locals distil from a mixture of cactus, other desert plants and seemingly dead rats. This fire-water they smuggle over the border to Pakistan, in milk-bottles bearing suspect labels - where *Johnnie Walker* was spelt Johny Walker.

Bombay was the next port of call - Here the challenge was to find a place where they rented cheap rooms. I negotiated a cheap three-day 'cash and no receipt' deal at about the rate they took for their 'hourly room rent' from the street ladies and boy-girls. During the days, I visited Elephant Island and ate at the small fresh food stalls that dot the city.

The sea breeze did not relieve the damp heat, which was clammy and oppressive, so I invested in a new safari shirt and cotton pants. My intent was to spend a few days in various five-star hotels, the ones with gigantic air-conditioned reception areas. Here I could pretend to be a normal US dollar 150 a day guest. The first hotel saw me washing dirty clothes in the gents, having a bath in the wash basin, followed by a decent shave and shampoo. All with unlimited hot water and where silky toilet paper was another luxury to savour. In addition, I could read the latest English newspapers and magazines and mingle with the American Express Gold-card holders. Here we swapped tales of our great travel adventures, and I sneakily-ate the sandwiches they left behind.

As midnight crept up, I went back to my bleak, windowless room, with condensation running down the walls, to the slamming of doors and eventually to fall asleep amidst stereophonic grunts of 'true passion' in the neighbouring rooms.

Fishing with the fishermen - From Bombay, the coastal ferry took me to Goa, where I slept on the open deck. Goa was once a prosperous Portuguese colony, boasting a number of impressive Christian churches, as well as Hindu, Jain and Buddhist temples. With the ambience of the Mediterranean, set against a backdrop of golden sandy beaches and easy access to marijuana, it was a true hippie paradise.

Twice I went out on one of the open-decked wooden fishing boats just before sunset. We pushed the heavy boats through the rolling surf to enter another world. Behind the white foam was an endless bible black sea, reflecting the stars, with erratic waves that allowed the moonbeams to dance. We cast nets and lines and hauled in fish of all shapes, sizes and colours.

Ganesh - The Elephant God stood proud on the bow of the boat. Ganesh is the Remover of Obstacles and of evil spirits: deity of wisdom and success. He is one of the most revered gods of the Hindu pantheon, and there are various versions as to how he came to have the head of an elephant.

One tells how the main Hindu God, Shiva, on returning home after a long journey, wished to enter the room of his beautiful wife, Parvati. He was stopped by a handsome young man. Consumed with jealousy, Shiva ripped off the young man's head, only to discover it was his own son, Ganesh. In great sorrow, he promised to replace his head with that of the next living being that came along. This was an elephant. He cut off its head and placed it on his son's body, so giving him eternal life - as the elephant god.

This is hardly less probable than the one night stand that resulted in our own Jesus being the son of God and his mother retaining her virginity.

Also improbable but true, these Hindu boatmen had a plastic statue of the Virgin Mary holding baby Jesus, at the other end of the boat. The crew sincerely believed in this faded Christian effigy: *'She our mother, with her we no drown.'* This suited me, as unlike her son I could not walk on water - my swimming ability was still akin to that of a concrete block.

One day there was a Christian wedding, a happy affair with deep southern-state gospel-style singing and hand clapping. Then one of the priests held up a flapping black cockerel and a white hen. The crescendo rose and rose as he waved the flapping birds around the church. Then at the speed of lightning he twisted their heads off and blood sprayed all over the bride and groom - and the congregation! A pulsating hodgepodge of gods, goddesses and voodoo magic.

It was time to head north, to Kathmandu, and then home to Oslo, where a new job, Sissel and baby Therese awaited me.

On reaching the chaos of early morning Delhi, I was greeted by a tuk-tuk driver who spoke perfect Oxford English. I was easily persuaded to hire him and his motorised three-wheeler, for the day and 'do the city sights'. Captain Jing Basantra turned out to be an ex-Army Officer and we were soon speeding around the city in his three-wheeler motorised rickshaw. We visited the usual tourist spots in both Old and New Delhi. He was a mine of useful and not so useful information, all colourful. We ate malai kofta (spiced bread balls in a rich sauce) and alu pakora (potato fritters) fresh from the street vendor stands and drank sugary milk tea with other tuk-tuk drivers.

Here he spoke about his exposing a corruption scam by high ranking Indian Army officers - where they bought land and then sold it to the army as shooting ranges at exorbitant prices. This exposé had not resulted in his advancement, but in his being court marshalled, found guilty and discharged, with subsequent loss of pension, house and family. He showed me the press clippings, and a copy of his last appeal to the High Court. The outcome he was still waiting for with some apprehension.

As the afternoon drew to a close, Jing suggested that I spend the night at Tuk-tuk Hotel. As we had been sightseeing all day and had not looked for a place to stay, I accepted easily. The hotel turned out to resemble a three storey car park, two floors were giant dormitories with about 100 beds in each. This was home to over 200 unmarried street sweepers, garbage collectors and rickshaw drivers - all men. A blanket and all their worldly goods were stored in a box or plastic bag under the bed, totally safe.

Below the two giant bedrooms was a cafeteria, resembling a work house out of a Charles Dickens' novel, which served all the residents. Here we ate generous portions of vegetable curry, rice and chapattis, with curd for desert. The other men were curious and shyly friendly as to my presence, yet they all showed a dignified respect for the other's limited space and private business. Jing insisted on my sleeping in his bed while he settled on a straw mattress by the side.

In a room full of 100 men, none of whom I knew, all trapped at the very bottom of society's ladder, I felt totally safe - and the lights went out at 22:00 hrs.

The next morning it was the same one dish menu for breakfast, with the option of a hardboiled egg. Once we had finished eating we did the not so normal tourist sites, including a leprosy centre and a visit to the Ghat where Gandhi had been cremated. That night we sat with the other hotel 'guests', drank *chiya* (strong, sugary milk tea), and sang songs accompanied by simple rhythmic handclaping.

It was touchingly sad to part from this fine man after three exciting days, all made worse when he thrust in my hands a Penguin copy of *'The Life and Times of Gandhi'* as he swung his tuk-tuk out into the traffic with a wave and a smile - and disappeared.

My next objective was Agra, to satisfy my touristic need to say, 'I have seen the Taj Mahal'. An awe-inspiring white marble mausoleum, built in 1632-53, by the Mughal emperor Shah Jahan in memory of his third wife. It overlooks the Jamuna River, made even more monumental when bathed in the light of a full moon. All I lacked was a bride, but there was no time to find one as the clock was ticking.

The city of Varanasi was the next stop. It was everything a holy city should be - even more so. I wandered the streets and viewed the ghats. Here the pyres of the dead seem to burn eternally, with the smoke transporting their souls to the next realm.

Living in low budget doss houses allows one a fleeting view into other people's lives. As does sitting with single men who do not have a kitchen to call their own, sharing and eating piping hot curries from the street vendors. Before sun rise or as the sun sets are the two times of the day when a village, town or city exposes their inner-selves to the casual visitor. Do not waste this time reading Lonely Planet or sleeping - get out in the back alleys and streets.

Journey's end - A rickety night bus took me towards the fabled city of Kathmandu. Here, three-and-a-half months after having left Europe, it was a jaded tourist who booked himself into the only US$1 room at the Kathmandu Guest House.

Here I was handed a heavily taped envelope by the receptionist, it was my share from the sale of the bus. The next day I bought a cheap ticket with a suspect airline, took a quick eight day trek to the Annapurna Base Camp and then flew home to Oslo.

NORWAY - Part 2

Chapter 14

Sports shop, mountain guide, trek leader and the Red Cross

Sissel and baby Therese, as well as Noble's job offer, had waited for me - even though I had arrived home six weeks late.

Running the outdoor department in a major retail shop in Oslo was another of life's schools. The company represented a dozen major sport equipment brands and decided to separate its retail business from its wholesale. I was to build up the outdoor retail department while Noble re-structured the wholesale business. A combination of managing the shop, holding climbing courses, writing for various magazines, and being a part-time social worker with youth in danger (drugs, etc.), as well as being both a husband and a father, functioned reasonably well, or so I thought.

Unfortunately, by the time Therese was four, Sissel was tired of the tempo I operated under so we decided to separate, a difficult process, but we continued to function as a family, living close to each other, and with both having equal parental responsibilities. Sissel had her boyfriends and I had mine (girlfriends), and we became the proverbial brother-sister combo - until she passed away nearly 25 years later of cancer.

Just after the divorce, a friend, Ulf Prytz, started an adventure travel company. Amongst other things, he had entered into a sponsorship deal with Arne Næss Jr, leader of the 1985 Norwegian Everest Expedition. The idea was that Ulf would organise trekking groups to visit the expedition at base camp. Here they would camp, meet the members and even go up to the start of the famous Khumbu Ice Fall. In return, Ulf would pay a small fee to support the expedition.

Ulf had intended to lead the groups himself, but with the business growing he was pushed for time, so he phoned me.

'Never led a group in my life; plus I have to look after Therese!'

'Don't worry, you've been to Nepal three times and the title Tour Leader gives you status and a 24-hour lead, you can read about the history and culture on the flight!' Adding, as an afterthought, ' - and take Therese with you.'

So I led three groups to Everest Base Camp. Therese, then eight years old, rode there on the back of her own mild mannered dzo, which is a cross between a yak and cow. We did her school work each evening in the tent and today she has a B.A. in Economics and a Masters in Innovation. A clear reflection as to my teaching skills!

Therese on Ricky - 'en route' to Everest Base Camp.   *Therese delivering Kransekake & Champagne to the expedition.*

The Red Cross - After seven years working in the shop, I spied a full-page newspaper advert: The Norwegian Red Cross needed staff and they needed them yesterday!

Norway's immigration laws were pretty naive and the authorities had no functioning infrastructure to cope with a flood of asylum seekers entering the country. They simply placed people seeking political asylum in rundown hotels in isolated areas around the country, with very few competent staff to run the 'centres'. As a result, split families, parentless children, seriously persecuted individuals, criminals and the insane were all clumped together, with little follow up. All waiting the outcome of their applications to remain in the country.

It was a ticking political time bomb, with the government hiding behind a ficticious 'immigration stop' policy. The Norwegian Red Cross, led by Odd Grann, saw this forced isolation as a humanitarian crisis. He took on the challenge, and converted M/S Fritjof Nansen into a floating reception centre, moored in the Oslo Fjord. Here asylum seekers received a health check, clean quarters, good food and information on their legal rights, in a secure and humanitarian environment.

I applied on impulse, had a lunch interview and got the job the same day. Our focus was centred on human integrity, welfare and the security of the asylum seekers. There were tales of war and woe, rape, murder, incest, drug running and horrific family separations.

Some were hardened criminals who quickly started dealing drugs and human trafficking from the safety of the ship. Quasi-marriages with Norwegian female drug addicts ensured asylum seekers could remain in Norway. The marriages were never consumated, and divorce followed a year later. The girls kept 10%, and the middlemen 90% - a lucrative business.

Our team established a Norwegian language school, kindergarten, medical centre, training room, shop and activity department. The majority of asylum seekers were ordinary people, looking for a better life - away from the hardships of their own countries. They established culture groups, and support groups for others in greater need than them. There were tales of heroism, of heart breaking losses, and of equally emotional reunions and of healing. There were hopes of starting anew.

In the middle of all this, Reidun, who had been on one of my trekking trips in Nepal, became my girlfriend. She was an island of peace in a hectic working day. Ten to fourteen hour work days, six days a week, would be relieved by late dinners, red wine and candle-lit bath tubs. In addition, we climbed together, and between us we created a fine healthy son. We never lived together, but Reidun was certainly the right mother for Filip - who, like Therese, provides a base for my life's satisfaction.

After two years of trial and error, the Norwegian government managed to establish a national system to receive asylum seekers. So they closed the ship despite the fact that we provided far better services and facilities than the government would, or could, in other centres. We had started the process of 'integration' - and had not simply ignored the hard realities of being an asylum seeker in a foreign country, as the politicians have done.

Not wanting to go back to the shop, I turned to freelance cultural projects while working part-time for Noble's sports agency, a combo that allowed me time to both climb and to travel. The first project was the production of a multicultural musical cassette for The Red Cross and Save the Children, to accompany four short stories about the life of child refugees. This 'teaching package' was then distributed to schools all over the country.

This led to organising a concert at the Oslo Concert Hall with 130 musicians and dancers from 12 countries: rehearsals, songs, instruments, contracts and music rights, sponsors, advertising, press releases, programmes, T-shirts, transport, security and accommodation to be managed. Things were going smoothly, which is always a sign of danger.

'Mandela is free!' - Turid, the leader of the South African Choir dropped the bombshell, *'My seven singers will now be thirty!'* I re-arranged the night's program, and forty hours later these joyful singers swung their way down the two aisles, the audience stood up, clapping and singing the refrains. The choir continued up onto the stage and the audience started to follow! Shit - then the other national groups came from backstage. I tried but failed to stop them. Two hundred joyous people were suddenly dancing, singing and clapping, hips gyrating on stage! A third song followed! I had lost control!

After three amazing years working with the Red Cross it was time to return full time to the outdoor sport industry. In the meantime Noble had sold his retail shop and was concentrating on wholesale. He had the agency rights for 47 major brands, from climbing to golf to fishing and tennis, with sales to over 500 shops. I picked up where we had left off, running the outdoor brands, but this time wholesale.

Freedom - Not only was Mandela free, but I too have known freedom - having had nearly four months free each year, with a full year's wage, for over forty years. This has allowed me to travel and to climb all over the world, usually on a shoestring budget. Often as a tour guide, member of an expedition or as a plain hitchhiking bum.

Memories flash back. Following the trail of Alexander the Great to the Oasis of Siwa (Egypt), where men marry men, to continue across the Great Eastern Desert, and marvelling at every twisted desert form. The feeling of sadness when standing before the Aral Sea (Uzbekistan), one of mankind's greatest ecological crimes - a shrunken lake, polluted beyond repair. Leading groups through Syria, Jordan and Lebanon and Morocco; slip-sliding alone along the mud paths of Stewart Island (New Zealand), expecting to be confronted by a Japanese soldier from the 2nd World War, wanting to surrender; trying to solo climb Mount Cook (failed); visiting Mother Teresa in Calcutta; bamboo rafting and climbing in Thailand; being a Holiday Inn tourist in Lhasa; re-visiting Muscat and Oman, where I had once been at war; snow-scooter exploration across Svalbard with Ulf; crossing Australia and eating 'croc-steaks' in Alice Springs; crossing U.S.A., trying to ascend Aconcagua (storm stopped play); to numerous solo ascents of snow-clad peaks like Pitz Rozeg, Pitz Palu and Pitz Bernina in the European Alps.

The Troll Wall - Europe's vertical mile overshadows the valley of Romsdal, a fjord-valley seven hours drive north of Oslo. It is a fortress of craggy edges, sweeping slabs, giant buttresses and, of course, The Wall. One thousand metres high!

1. Troll Pillar (1958). 2. Swedish Route (1978). 3. English Route (1965).
4. Fiva Route (1931). NB: Route lines are approximate, drawn in Kathmandu nearly 30 years after climbing them. Romsdal guide book recomended.

The original Fiva Route had first been climbed by two cousins, Arne Randers and Erik Heen, in 1931. Ken Wilson, editor of Mountain Magazine, asked me to interview Arne - who was both mountaineer and a war hero.

'We knew that Noel Odell, the Himalayan mountaineer, had climbed solo up the obvious vegetated break behind Fiva Farm. We had some basic equipment discarded by English climbers, with this we climbed some interesting mountains. Then we decided to try Odell's line. At the hanging glacier, we put on the rope, but there was nowhere to secure it, so if one fell he would pull the other with him. That is how we thought people climbed. After one hard section I called for Erik; as he moved, the whole block fell away.'

Darkness came, so they lay on a small ledge and went to sleep. *'When we awoke we were covered with snow, which we brushed off and continued.'*

Fifty years after their ascent, I climbed the same route on my own to try to gain some impression as to how it must have felt for these two young country cousins back in 1931. I had a guidebook, lots of experience and a sleeping bag, whereas they had little experience, just the genuine drive for the unknown.

What an incredible undertaking, powered by the enthusiasm of their youth and, it is still a serious route even with today's equipment.

The next development was the ascent of the Troll Pillar (1958), said to be the longest climb in Europe. Again by Arne, this time accompanied by the younger Ralph Høibak. Long it is, but overhyped as a historical breakthrough, not being comparable in quality or difficulty to what had been done in the European Alps. Only two and a half pitches require a rope for a competent party - more a steep scramble up a very vegetated ridge than a route of quality. The pillar does mark a border, for between it and Fiva lays the Troll Wall itself - Europe's Vertical Mile, capped with a Manhattan like skyline.

Here the classic English Route, as the name suggests, was first climbed by a British team, in 1965. This complex line weaves its way up a very impressive rock landscape and it was the most popular route up The Wall until 2006, when a giant rock fall took out a big section of the route. The less 'classic', and rarely climbed, Norwegian Route was climbed at the same time. The popular press turned these two ascents into a race between the two nations - which it never was.

Me somewhere on The English Route. 'Who tilted the camera?'   *5-Star accomodation - shower included.*

Most photos show The Wall as being wet, damp and running with water. As such, one assumes it would be easy to find water - wrong! To save weight Steve (Helmore) and I took only two 3/4 litres bottles of water on the English Route. By the end of day two we were reduced to licking damp rock and sucking the dew off evil-tasting lichen! Experienced climbers - ha!

The Hilton - The Swedish Route was first climbed by two Swedish climbers, Johansson and Nilsson, in 1978. They created a masterpiece that took an almost direct line up and through giant overhangs and free hanging pillars of hostile rock. We intended to climb it free, in two days, and found the climbing steep, varied, very sustained and interesting. Plus the rock on the whole was surprisingly solid.

By early evening on day two, we had reached the Hilton Ledge, which was just 100m below the top, our goal was within easy reach. But Ted (Hall) refused to continue. *'What a bloody brilliant place - we have to stay here - no wonder they called it The Hilton!'*

Below us were over twenty pitches of grade VI and VII free climbing that had taken us through an impressive world of over hanging rock. We were so close to the top, but he was right, the Hilton was the place to be. We lay perched on two small body-wide ledges, eating goulash soup accompanied by red wine. For dessert, was a 1km high free sweeping wall of cold satanic grey freckled Troll granite.

It was like standing on the bow of the Titanic, although three things were missing: music, an iceberg and a beautiful woman.

We awoke to thick pea soup mist and drizzle the next morning, with only two short pitches left. We reached the top wet, but happy.

Three accidents in fifty years of climbing - The first accident had been that idiotic 20m fall off Yob Route, resulting in three months in hospital. The second stupid mistake, again totally my fault, was my fumbling, that resulted in my gloves tumbling 300m to the glacier below on Store Austbotntind, in the Jotunheimen (Norwegian Alps). We were attempting a first winter ascent on a twenty pitch route, up the SE Ridge. Jarl (Saetre) was 40m above me and Ben (Campbell-Kelly) was hauling bags below. The extra gloves were in one of the bags below! By the time Jarl had descended to me and we had abseiled down to Ben to get the gloves it was too late - my fingers were frozen stiff. This stupid act of mine resulted in a painful descent, down frozen ropes, to reach an exposed sloping ledge where they set up our tent in the approaching darkness.

Normally, if one freezes one's fingers or toes it is best to reach medical help before defrosting them, so you don't damage the skin and flesh further. However, the thought of lying awake for ten hours with deep frozen fingers was not very appealing. We knew I could descend and virtually walk out (in knee deep snow) the next morning, to reach the nearest road with my hands protected in warm down booties. As such, we felt the risk of defrosting was acceptable.

Jarl started cooking my fingers medium-rare, while Ben kept the tent from collapsing under the weight of the continuous snowfall. It was a painful night full of self-incriminating thoughts. Not so much about the risk of losing fingers, but the inconvenience it caused the others and how embarrassing it was to be an idiot.

Next morning witnessed a trudge through waist deep snow, rather like Napoleon's retreat from Moscow - we only lacked dead horses and abandoned cannons.

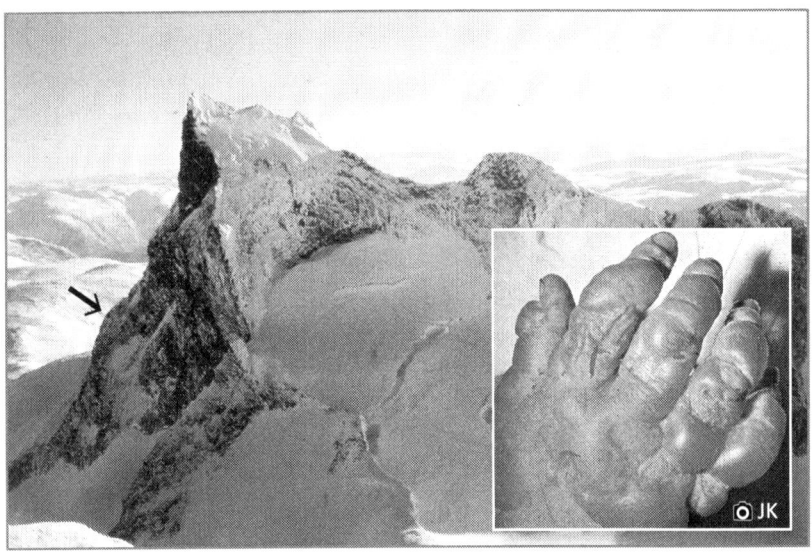

Store Austbotntind. ↘ *Point from where I dropped my gloves. Luckily, I lost no fingers. The moral is: Don't drop your gloves and do carry spare ones!*

The third incident was not an accident - More an Act of God, that took place on one of the great classic climbs in the Italian Dolomites. One summer, around 1980, Keith (Robson) and I had climbed a dozen grade VI and VII routes on the Sella and Vajolet Towers - fast and in good style (using no aid). With only three days left, we drove like mad to reach the Tre Cime - a magnificent trio of Dolomite peaks. Here we chose the *Yellow Edge*, which we climbed fast, free and in the rain. Early the next morning we started up the *Comici Route* on the North Face, again in the rain.

The first 300m presented varied and sustained grade VI and VII-climbing. We planned to climb it free, but it proved too sustained for me so I used three pitons as handholds, whereas Keith did the whole route free. The wall is slightly overhanging, so we were protected from the rain and as such, remained pretty dry. The climbing was hard and complex, yet on excellent rock.

Comici Route, VII-, Tre Cime, Dolomites.   *Keith, lunch, halfway ledge.*

On reaching the halfway ledge, we were now exposed to the drizzle, plus there was a 'warm electric' feeling in the air. Obviously, a storm was building up. We ate our sandwiches before starting to climb up the V-groove towering above us.

Keith led the first pitch, and as I followed, about 50m up, we were suddenly engulfed in a banshee wave of electrical energy.

Stunned - hit by lightning! (See arrow)

Not funny, not happening to me, not hurt, just numb, light headed, surprised and feeling nauseous. Had I passed out - yes or no? Time passed, one minute or ten? I was dangling from an old piton, paralysed down one side, screaming, *'I've been hit!'*

A lifetime later, Keith shouted back, *'I've been hit.'*

'So what, I've been hit twice!!! Take in the fucking rope!!!'

Keith took in the rope as I slowly and painfully climbed up to him. Here we sat, grey faced, wet and scared, with neither of us knowing whether we had passed out for one or ten minutes. His feet burned, and weeks later his toenails fell off - roasted by the electric current. The way down would mean complex abseils, with stiff, wet ropes that might get stuck. A single error and one of us could be left hanging way out from the rock, too exhausted to jumar back up to regain the rock face. The air was full of wet, warm electricity. There was nothing to discuss, as the only way out was up, over the top and down the normal route to the valley below.

Progress was slow, on holds covered with slush, while the air all around was clammy warm. We felt the electricity in our teeth. Keith, the better climber, led most of the pitches. Trying to keep out of the groove, taking the more difficult and poorly protected right-hand wall where we naively felt the current would be less powerful.

As we climbed I struck a deal with God, and promised to be a better Christian in the future. Three painful hours later, we both stood on the top (not God and I, but Keith and I) in ankle deep new snow, just as the sun went down, and my promises to God were soon forgotten.

Me under Austabotntind, 2204m, Jotunheimen - Norway's Alps.
Photo: Keith Robson.

Book 2

GLIMPSES OF NEPAL

Lama Ap Thakpa after attaining a spiritual trance. Badhe Festival, Manang, (1979). Photo: Zdeněk Thoma.

The Federal Democratic Republic of Nepal is a landlocked sovereign state in South Asia. With an area of 147,181 square kilometres and a population of approximately 27 million. The world's 93rd largest country by land mass and the 41st most populous. Located in the Himalayas and bordered to the north by China, and to the south, east, and west by India. Kathmandu is the nation's capital.

Contains eight of the world's fourteen 8000m peaks (see map), and more than 240 peaks over 6000m.

Hinduism is practised by about 80%, Buddhism by 10%, Islam by 4%, other religions include Kirat, Christians and Animism.

Chapter 1

To Trek

The word trekking originates from the language of the Boers in South Africa. With British domination in the area, the Boer or Dutch farmers, packed their belongings into wagons and, together with their families, livestock and slaves, started The Great Trek. This took them deep into the hinterland in search of new territories. Today, to trek means to hike or to walk, usually carrying your own equipment for a number of days or weeks.

Trekking as we know it has its roots in the early mountaineering expeditions. They took steamships from Europe via the Suez Canal to Bombay, to be confronted by customs officials and reams of red tape. They then crossed the subcontinent by train towards the border of Nepal. From here, with hundreds of porters carrying all their equipment, they started the long trek to the base camp of their chosen mountain. The enormity of the undertaking resulted in Colonel Jimmy Roberts (1916-1997) offering a service that covered everything from peak permits to importing of equipment. In addition food had to be purchased, packed and transported. To do this, a great number of porters were hired to ensure that everything arrived on time, in good condition and without any losses.

Base camps had to be organised and a supply chain up the mountain established. Here the climbing sherpas kicked in, but they too had to be trained and equipped. The western climbers usually chose the route, did the climbing and secured both the ropes and the ladders. While the sherpas did the carrying of equipment, fuel and food once the route was opened.

Often there was a sponsor, a wife, or a group of friends who joined the expedition as far as base camp. Some provided important and legitimate finance, as they undertook genuine research. Parallel, and a major influence for what was to become the organised trekking industry, were the botanists, geologists, anthropologists and other special interest groups. All these needed a guide, cook and a complete support system for the walk in and the return journey.

Jimmy Roberts, an ex-military man and mountaineer, was ideally suited to the task of providing such services. He had spent much of his army career in India, from where he had gazed longingly at Machhapuchhre (Fish Tail Mountain). He led an expedition that climbed the mountain in 1956. The classic *Climbing the Fish's Tail* by Wilfrid Noyce has been re-printed and offers an exciting insight into the obstacles the early pioneers faced. They also opened the route into the now popular Annapurna Sanctuary. It was Roberts who later persuaded the government to officially acknowledge Machhapuchhre as a sacred mountain, so prohibiting further climbing of the peak. In addition he obtained governmental permission for organised climbing on the lower mountains, which we now call Trekking Peaks.*

Roberts is without a doubt the *Father of trekking* in Nepal. Mike Cheney (1928-1988) was also at the forefront of this new tourist industry. He established a number of trekking agencies and empowered Sherpas to own and run their own companies. Mike, a former British officer, was an unusually outspoken and flamboyant individual, often seen cycling the streets of Kathmandu in a kilt!

Then there was Odessa-born Boris Nikolayevich Lisanevich (1905-1985). He was a traveller, ballet dancer, big game hunter, friend of royalty and the famous, and above all, hotelier extraordinaire. He converted an old Rana Palace into the Royal Hotel and later established the world famous Yak and Yeti Bar and Chimney Room. Boris developed the upmarket tourist industry and his colourful life is vividly presented in *Tiger for Breakfast* by Michel Peissel.

The Overlanders - They crossed between Europe and Asia in battered Volkswagen minibuses or Beetles, or as part of commercial groups in refurbished coaches or heavy-duty trucks. These were mainly experienced and hardy travellers who, on reaching Nepal, adjusted easily to staying in local houses and eating local food along the mountain trails.

Overland travel, while the dream of many was in reality the preserve of the few, being both time-consuming and expensive. Many who did make the journey continued on to South East Asia, Indonesia and Australia, as trekking was not their goal.

As late as 1977, on my first visit, trekking was restricted mainly to the Solu Khumbu region, as well as the Helambu-Langtang areas or the southern and western sides of the Annapurna range.

* *Recommended: The Trekking Peaks of Nepal by Bill O'Connor.*

In Kathmandu the majority of travellers who had journeyed overland, including the hippies, stayed in Freak Street near the temples of Durbar Square. Here were *hash cafes'* and *happy bars* where one could sample the goods. Seriously, you could actually try the hash and other offerings on the 'menu' before buying!

'Apple-pie surprise' had a crispy crust heavily garnished with marijuana, while 'Flying Omelettes' were laced with red and green peppers and hallucinating mushrooms.

Central Hashish Store was a popular meeting place for hippies and bikers. Sale of hashish was banned in 1974.

Kathmandu Guest House - became a heaven for overlanders and mountaineers.

Eating the latter resulted in one hippie-friend saying he wanted to fly home, to which, and to our surprise, he started clucking like a chicken, flapped his arms and promptly flew off the flat-roofed second-floor café. This resulted in a broken arm, a fractured jaw, and the loss of five teeth. You have been warned.

Freak Street's position as the tourist centre was challenged in 1968 when a young botanist, Karna Sakya, converted one of the old Rana palaces into the Kathmandu Guest House (KGH), in Thamel. This move was followed by KC, a young pot smoking entrepreneurial Nepali, who established his popular restaurant in the same area. He also organised the Ricksaw Chariot Race, a wild competition between Nepali and tourists. The travellers followed.

The trekking industry - This did not explode in terms of numbers until the advent of regular and affordable air travel. Between 1950 and 1980, few of the facilities that are taken for granted today were available. It was either expedition style trekking with full porter, tent and kitchen service, or you ate and slept in local houses.

Money earned from trekking and expeditions by the Sherpas of the Solu Khumbu district was invested in establishing lodges and their children's education. The Sherpa community originally came from eastern Tibet, about 500 years ago, and they settled in the Khumbu district, under the shadow of Mt. Everest. Their name roughly translated, being, *'the people from the east.'*

Many traditional sherpa houses are solid affairs, built of rough hewn stone and roofed with wooden boards or slate slabs. On the ground floor, the animals are quartered at night and the floor above consists of a large living room and kitchen combined. Here, an open fire forms the heart of the home. In the past, this rarely sported a chimney, based on the belief that evil spirits could come down the chimney and enter the body while they slept.

Typical Sherpa houses. Some houses have a small altar, others have their own prayer room artistically adorned with effigies of Buddha and various deities.

Windows were without glass, hence they were small, or covered by wooden shutters to stop the wind penetrating, and as such, they allowed in very little light. By making cotton wicks from strips of old T-shirts and sticking them in used gas cartridges filled with oil, a dim light helped guide us through the evening meal. In the old days, there were no toilets and when nature called in the depth of night, we often had to climb over the family livestock to get out of the house.

Everyone, family and guests alike, sat around the smoky fire to keep warm, to cook and eat food and to discuss the day's events and to plan tomorrow's. The smoke both cured hanging meat and naturally impregnated the wooden roof against rain. Some houses had separate sleeping quarters, while in others quilts would be taken out of giant boxes and laid down on the floor.

Often, three generations would sleep in the same room, with the act of reproduction being undertaken as discreetly as possible, to the just as discreet giggles of the 'sleeping' older children. It was not uncommon for a woman to give birth by the warmth of the kitchen fire, whilst grandmother cooked the family dinner and father was downstairs milking the naks (female yak).

With the development of tourism, partition walls were set up to provide bedrooms, yet still, we ate with the family around the open kitchen fire. Sherpa hospitality became a by word, the industry grew and new lodges were built. These had glass in the windows, sheets of corrugated iron on the roof and wood burning stoves with back boilers to provide hot water. Toilets and chimney pipes became the norm, the latter sporting a small chimney roof to keep out the rain, wind and of course, the ever present evil spirits. Electricity followed, often via small local hydroelectric power stations.

In the tourist season, spring or autumn, the wife would normally run the lodge while ama, the grandmother, looked after the babies. The husband would, when work was available, join an expedition or go trekking as a porter or guide.

Everest was the goal for trekkers and climbers alike, the magnet. Yet, a major factor in the rapid development of the Solu Khumbu area was Sir Edmund Hillary's organisation, The Himalayan Trust. Hillary was well aware of the role the Sherpas had played in his ascent of Everest, and to the development of Himalayan mountaineering. He wanted to contribute something to the people and to the area. He started by building schools, progressing to medical aid posts and forest nurseries, and scholarships. The airport at Lukla was built in 1964 and through this gateway, foreign tourists and aid workers flooded the area.

Within twenty years nearly every village had its own trekker's lodge or hotel. These were followed by shops offering everything from western food to advanced mountaineering equipment. Soon bakeries, bars, pool halls, cinemas and cyber cafes popped up like mushrooms. A similar development pattern has occurred as tourism spread to other areas of the country.

Chapter 2

Nepal calling

With nearly forty visits to Nepal over a thirty-five year span, I have been fortunate to have entered and exited the Solu Khumbu (Everest) region from the west, the south and from the east. Walking virtually every path and crossing every pass in the region, as well as climbing a few mountains along the way, with friends, or leading organised groups, or on my own.

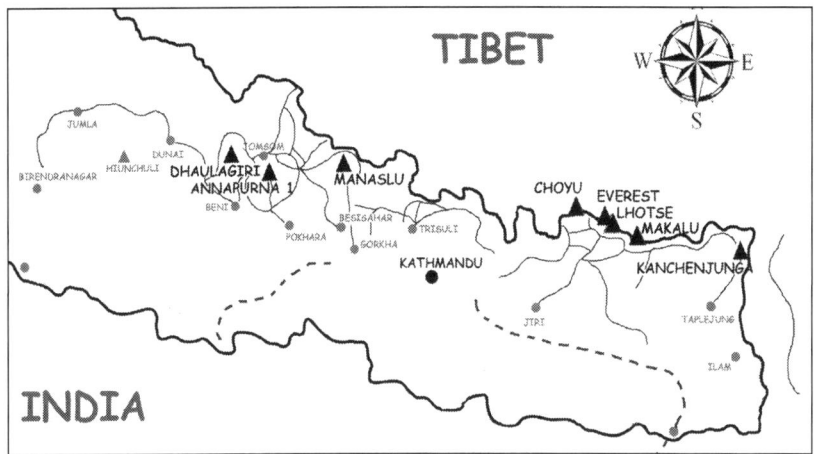

Black line = My trekking paths, over the years. Dotted line = River rafting

Other journeys have gone via the district of Rolwaling, crossing Tashi Labsta La and climbing Parchomo, 6273m, as a side trip. As well as, exploring Helambu and Langtang - and trekking towards the seldom visited Ganesh Himal and further around the majestic Manaslu. Even after circumambulating the popular Annapurna Range over a dozen times it still has a magnetic attraction. This says something about the majestic mountains and the people who live in their shadows. Yes, roads have brought the unfit and unworthy to Manang, but the mountains remain the same and there are many 'diversions' that still make the 'circuit' worthwhile. The upper path from Pisang for example, or crossing over to Jomson via my favourite: Tilicho Lake.

Wonderful side trips have taken me to the popular Annapurna Sanctuary on the Pokhara side, also to the original and much more demanding French Base Camp (on the western side of the range), and on four occasions to the once forbidden region of Nar-Phu.

A tree at this altitude takes 30 years to reach maturity. Photo: J. Haglund.

The first two times, it was like stepping back 300 years in time. Unfortunately, fifteen years later we witness the result of uncontrolled cutting down of the sparse woodland for the building of tourist lodges. Heralding a new wave of tourists bent on discovering yet another 'last great forbidden kingdom' before it too disappears.

It seems new tourist areas do not learn from what we (yes, we are all guilty) have done wrong, time and time again. If local entrepreneurs organised a kerosene depot, and/or a small hydroelectric plant, and we the tourists were willing to pay a small surcharge, then we could reduce this onslaught on the fragile forests - and slow down the inevitable erosion that follows in our path.

We could also reduce our expectations with our incessant demands for hot showers and warmed dining rooms. What's wrong with an afternoon wash in a clear mountain stream and the wearing of a down jacket in the evening? Wood consumption could also be reduced by ordering the same food as other trekkers.

Yet, I am a little unfair, for tourism has led to increased income and higher living standards for countless thousands, which in turn has led to increased use of timber for house and school buildings and not just for lodges.

The eastern flanks of Annapurna have also appealed, with two unsuccessful attempts to traverse from Sikhles towards Lamjung Himal, across the Namuna Bhanjyang La, 5560m, from south to north to reach the Annapurna Circuit. The first time I was stopped by heavy mist and snow, resulting in my not being able to find the pass! The second attempt was stopped by rock hard snow at the pass that meant crossing without crampons would be too risky for heavily laden porters.

Eventually, on the third attempt, I was rewarded with a complete crossing of Lamjung Himal, a gem of a trek, starting in the north and ending in Pokhara. Accompanied by three porters, after having acclimatised in the Nar-Phu area, I used ten days and only got 'slightly' lost once. Tempted? Then do take enough kerosene, make sure your porters are well equipped, do not burn vegetation, and take your litter out!

Chorten: *The form represents Lord Buddha sitting in a meditative position on a throne, often of water lilies. The oldest form of Buddhist monument, five main forms each depicting different functions. Some contain relics of holymen, others items owned by Buddha/his disciples. Others symbolise aspects of Buddhist theology, others mark holy places and festivals.*

Another time capsule was visiting the once mysterious Kingdom of Mustang, to explore the small temples and burial caves and catacombs of a long-forgotten civilisation. Here, 1800 years ago, bodies were stripped of their flesh, so that the skeletons themselves would not disintegrate as the flesh rotted. The skeleton was then placed in a wooden coffin and preserved in a sitting position. Recent discoveries of mummified remains and their possessions, including gold-plated burial face masks, if not destroyed by ineptitude and greed, may see Mustang developing into the Egypt of the Himalaya.

The charm of this area has been considerably reduced with the completion of a road to the Tibetan border, with the jeeps and trucks that ploy this once peaceful area. An increasing trend, as roads reach far into areas that were once remote mountain regions.

Writing about past walks evokes a thousand memories. One in particular, at the Dhaulagiri Base Camp, when Slovenian climber, Tomaź Humar, kindly offered to take a member of my group, who had a serious bronchial infection, out in his expedition helicopter. As we put her on board, he turned and smiled, *'Cost you a beer in Kathmandu, Mr. Tourist Leader!'*

Ten days later, his porters deserted him, as the approach to the South Face of Dhaulagiri had proven too difficult and dangerous. He was stuck with all his gear, and as we were near the end of our trek, I could let him take over my experienced porters, to carry his loads.

We tourists may not make history, but it was exciting to wish him good luck on his attempt to climb this 4000m face, on his own, climbing at night, when the stone fall and avalanche danger was less.

He succeeded - a historical event.

Another flashback, was the night when the heavens opened up and seemingly hundreds of hailstones ripped through my tent like machine gun bullets. A lonely experience indeed, with three days to the nearest village.

Writing brings back faces, names and places - as well as trails. Like a thirty day walk with friend and guide Lakpa Bhote through the middle hill regions of west Nepal, then crossing Lower Dolpo to reach the town of Beni. We only met two other tourists along the way, a magical journey, buying food whenever we found a village or a shepherd's hut. One savours living in local houses and eating with a family, with no coffee, no Internet and no TV. They treat you as honoured guests, and you become the evening's entertainment. Even when a village has electricity, each house may only have one bulb, of which they are proud. By its dim light, the mother cooks the evening meal while the children do their homework. Yes, there are still alternatives to the klondike disco towns of Namche and Manang.

Memories flood back of torrential rain soaked paths on the way to Kangchenjunga, the third highest peak in the world. Sleeping under a hastily constructed leaky bamboo roof with Lepcha porters and sharing their simple food. A few days later, repeating the experience, this time in a lone yak-herders makeshift abode - full of smoke, nak tea, potatoes in their skins and lashings of humour.

The crossings of more than thirty different snow-covered passes over 5000 metres are stored in the memory bank.

A memory bank - from which one can always draw, yet never deplete. Muscle testing days to reach a pass, where one is then able to sit and gaze upon unnamed summits - some climbed, some not. Time to climb their ridges in one's mind, imagining yourself in a lonely tent perched on that far arête, with just 600 metres to the summit. Maybe in the next life, who knows? Peaks upon peaks, some draped in cloud, others dazzling white under a diamond blue sky.

Finding a snow leopard's footprint, giving succour to a dying man or passing a twisted scarlet fungi growing from a rotting tree trunk - all add to the kaleidoscopic journey of life.

Once I crossed the path the great botanist/explorer Joseph Hooker had followed in 1848. Hooker, a close friend of Charles Darwin, had been dependent on local yak herders to guide him across this then unchartered territory. Whereas, I had a map and an occasional Buddhist flag to confirm the way, as we worked our way through the new snow and dense mist. Such days tend to run into weeks and time stands still. Inconveniently, for my western mind-set and ego, such journeys reveal that our own passage through this Himalaya landscape is of no consequence to anyone other than to ourselves.

Another time, accompanied by Bhote porters (a cast of Tibetan origin, who like the Sherpa moved into Nepal, with the majority settling in the Makalu area), I was confronted by the anarchy of the Maoist uprising. My plan was to cross from Kanchanjunga to Makalu, and climb a peak on the way.

During the armed conflict the government had lost control in many areas so that TIM cards, which are tourist trekking permits, were not being issued. Feeling confident that there would be no police or permit control posts in Maoist territory, I set off. Ten days later I was stopped at a Maoist control post, red bandoleers, guns and all!

'You pay NRs. 50,000. We give permit.'
'I don't want to buy the mountain! NRs. 3000.'
'Too little, you no go!'
'I am going - are you going to shoot me?'
They discussed this option between themselves.
'We don't like shoot you. Cheap permit, NRs. 20,000.'

Not wishing to create unpleasantness for my porters, I agreed to NRs. 5000. He wrote out a Maoist Party monopoly type pass, which granted us permission to follow the almost non-existent trails along the Nepal-Tibet border for another fifteen days. We struggled through deep forested valleys, crossed wide open grazing meadows, and passed below more than 20 peaks, seemingly all without names and probably none having been climbed, to reach the Makalu region.

Years earlier - With three different Bhotia porters, I had crossed from this area, over the Sherpani Col (6180m), the West Col (6190m) and the Minbo La (5845m) to reach Solu Khumbu. This is probably the most demanding high altitude trek in Nepal, if not in the whole Himalaya. We four carried everything: food, stoves, fuel, tents and climbing equipment. Every ounce had to be saved, nothing could be forgotten, for once we left the last village, we were on our own.

I cut down to two pairs of underpants, two pairs of socks, two t-shirts and one of everything else. We ate the same food, dal bhat (rice and lentils) twice a day, with the occasional potato or carrot thrown in. My only luxury was a solitary morning cup of coffee. Theirs was sugar in their tea. We had one 40mtr 8mm rope, two light ice axes, three aluminium snow stakes, one ice screw, a few slings and, to save weight, just one crampon each. Only Lhakpa had previously abseiled, so, on a wind swept Sherpani Col, I held a crash course in abseiling - the gentle art of sliding down a rope in a controlled manner.

We moved quickly across the glacier, intending to cross the second and more daunting West Col (pass) the same day.

By late afternoon I had stamped down the candy floss snow, put in two snow-stakes and sat on them, giving the porters a safe top rope down the first 40m. Here they kicked out a ledge and put in the third stake, tied themselves to it and waited with that 'doggy-eye' trusting look. If I were to fall while connected to them, it would see their snow stake ripping out and all of us cartwheeling to the glacier with a loud thud. So, I let my end of the rope slither down to them and climbed down without it.

We repeated the process and quickly lost height, to find the last 15m section slightly overhanging. Here, as evening darkness set in, I sat on the three snow-stakes, first lowering the baggage and then them, one by one. Their dark faces were deadly serious as they went over the edge. Then it was my turn. Without someone sitting on the snow-stakes, I hoped the last fourteen days sparse food rations had reduced my weight. I abseiled out into the void, and could just make out the neatly laid sacks, sleeping bags and mats ready to absorb my fall. All three were looking up, arms outstretched to catch me.

We camped on the glacier, and reached Chukung in time for a mammoth breakfast and real coffee.

Chapter 3

The Annapurna Circuit - aspects of village life

Behind the Annapurna's are the twin villages of Pisang. On my first visit thirty years ago, there were only four buildings in the lower village and Upper Pisang looked down upon us with a foreboding medieval aura. Not surprising, when realising that the area had been closed to the outside world for hundreds of years.

Ten years after this first visit, Jon Tengs, a Norwegian student, hired me to guide him up a Himalayan mountain. His requirement being, *'Any mountain, will do.'* To finance this dream he sold his father's second car (not sure his father knew about this until after the sale). We left Norway with Aeroflot (cheapest airline), as a car held together with paint and string had not commanded a premium price. The outstanding debt, Jon promised to cover when we returned home, by selling his classic collection of 120 VHS Bruce Lee type adventure films. I should have smelt a rat; and realised that such films would have a limited market appeal. Years later, Jon cleared his debt, and I ended up owning this 'priceless' collection!

Pisang Peak, 6091m - This peak was our goal and an ideal introduction for Jon. There was an eight hour bus journey, followed by a ten day gradual walk in from the hot lowlands to reach an alpine landscape of juniper and pine trees. There is no glacier, nor steep approach, nor is the climbing difficult.

On reaching the village of Upper Pisang, we discovered there were no lodges, but soon found lodgings with two sisters. The plan was to use a few days to acclimatise, and then move our equipment up to a small base camp. On arriving here, we were hit by a snow storm, so we descended to the flesh pots of Manang to buy more food. Revived by beer and apple pie, we returned to base camp - a three day round trip. Next morning, with the snow fast melting, we left our porters and established a bivi high camp. The next day, we ascended about 400m, to the snow line, and then descended to spend a further night, to aclimatise before making an attempt at the peak itself.

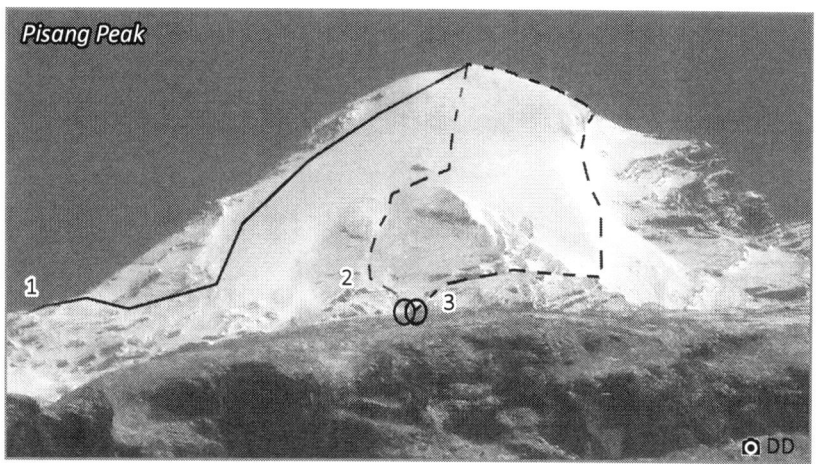

1. Normal Route. New routes: 2. South Face Direct. 3. Moon Walk.
⦿ Bivouc. I tried but failed to climb the far right hand edge.

Up early, we retraced yesterday's route, along an easy but exposed rocky ridge, this time guided by our headlamps. On reaching the snow slope, in half light, we sat down for breakfast. Above us was a long snow field, which capped the whole mountain, and after two hours of easy snow climbing we reached a point just below the summit itself. From here, there ran a knife thin ridge, to the real summit, just a few metres higher. The risk of the snow fin collapsing under our weight was not worth the risk, to gain a few metres.

Instead, we feasted our senses on a magnificent 360° Himalayan panorama. Manaslu to the east assailed our eyes, as did the Chullu's to the west, the Tibetan plateau to the north, and the whole Annapurna range to the south - paradise unsurpassed.

We then descended to the beer bars of Manang. From here we sent our porters back to Kathmandu and continued to the Thorong La pass, at 5416m. Here we shook hands and parted. Jon descended to Jomsom and continued to Kathmandu. I camped just below the glacier and the next day climbed Thorong Peak, 6144m in three hours, and glissaded down in 30 minutes. Now I had a choice: return home or do another climb. I headed back to Pisang.

Once there, I took a sleeping bag, stove, fuel and food, crampons and two ice axes, and spent about seven days on the mountain. I first tried the long rock ridge on the eastern side, encountering steep and at times rotten rock. Finally, I admitted defeat and retreated, to spend long, lonely hours, cooking freeze dried beef stroganoff and talking to myself. Questioning the meaning of life, but never questioning why I was sitting on a little ledge all alone, high in the Himalaya.

The next day, I scrambled and then climbed a line, at about grade III/IV-, up some rocky slabs and buttresses. This I baptised Kellogg's Wall, which reflected the quality of the rock. 300m of climbing lead to a ledge system just below the main snow field. These ledges provided the ideal bivouac site - safe, comfortable and with incredible views. Here, I spent three days and climbed two new routes - one up the middle of the face. The other line went up the right hand side of the south face. The climbing was on hard-packed snow, which afforded sheer pleasure, while the whole Annapurna range burnt itself into my back. An office with a view!

Hospitality - Over the years I would return to visit the sisters in Pisang and deliver some photos. On one visit Kalpana, the elder, gave me a welcoming smile and the words, *'Come, come...'* as she started preparing Tibetan tea. As is the normal practice, her neighbour soon joined us and we sat around the dark, smoky kitchen cum living room chatting away and swapping stories.

People came in and out, a baby suckled a mother's breast, a grandmother thumbed her prayer beads and a child was asleep on her sister's lap. A young ani, a Buddhist nun, in a worn maroon robe, beads in hands, lips gently reciting her prayers, could have easily graced the catwalks of Paris, London or New York.

As decorum decrees, I drank the offered tea. This is not Lipton's Yellow Label, nor Twining's Earl Grey, but Tibetan tea.

Churning Tibetan tea

Tibetan tea - This originated in the 10th century and gained great cultural and economic importance by the 13th, under the Phagmodu kings of Tibet. From here, it spread in popularity all along the Himalayan regions from Bhutan to Kashmir. There are various ways to prepare it. One is to boil the large black tea leaves in water for half a day. This is then skimmed and poured into a long wooden cylinder. Then nak (female yak) milk or butter is added and some salt completes the brew. It is then churned to produce a dark brown liquid with the consistency of a thick broth.

Before being served, it is transferred into a kettle and heated on the kitchen fire. When ready, it is poured into wide shallow cups or small bowls. Here a handful of Tsampa flour is often added to the bowl, turning the tea into a thin porridge. Tsampa is roasted barley that is stone ground into rough flour. It is slightly sweet in flavour, has a creamy texture and is packed with nourishment. In addition, the flour is also used for making Tibetan bread and chapattis.

Before starting the day's work it is not uncommon for a person to drink half a dozen cups of Tibetan tea, which provides the main sustenance for the day.

Tibetan tea drinking also has a strong social aspect, bringing people together for hours on end to discuss village life.

As in Japan and China, there are strict tea rituals in Tibet and in Nepal. Should one receive guests, then custom demands, that they drink the tea with a little pause of appreciation between each sip. The guest is expected to drink three cups. This is difficult because time honoured etiquette requires that when the first cup is half empty, the host fills it up to the brim again. Followed by the dreaded words, *'Shay shay, shay shay'* - meaning: More more, more more.

To the uninitiated, Tibetan tea tastes rather like a mixture of old tractor oil, sawdust and melted-down rubber tyres. Such never-ending sessions require a strong will and an asbestos stomach.

You have been warned!

Yak cheese - The yak (male) and nak (female) are long haired, broad shouldered members of the bovine family. They are tough, hard working beasts used to freight goods, plough the fields and provide much needed dung for manure and for fueling cooking stoves. They are an important source of meat, rough wool and leather, thus they play a central role in the life of the Himalayan peoples. In addition, nak milk is used for making butter, cheese, as oil for cooking and for fueling lamps.

There is a misunderstanding regarding the gender of the beast. A yak is the male of the species, i.e. a bull. Whereas, as noted above, the nak is the female, and produces milk, from which they produce cheese, called chhurpi.

Therefore, it is nak cheese one should ask for, **not** yak cheese.

The cow is holy to Hindus, as she symbolises Mother Earth - one that eternally gives while asking for nothing in return. Hinduism is the main religion of Nepal, and before the Maoists took over, to kill a cow was illegal. The people of the high mountains are predominantly Buddhists and they are not supposed to kill any living creature.

So the slaughtering of a yak or a nak puts them in a quandary. Many of the Sherpa caste pay other lower castes to do this work, having no qualms about eating the meat as long as they do not have to slaughter the animal.

In many areas, butchers are regarded as low caste, and untouchables. In Tibet, they have their own living areas outside Lhasa, the capital. However, the Nepalese are a tolerant, pragmatic people who like their meat, which in this cold, is a necessity for many. So out of legal, moral and religious considerations, a yak or nak is never slaughtered, it has a 'pre-arranged accident'.

The accident - This begins when the yak or nak is led to a field behind the house. Its feet are tethered together and whipped from under it, so it falls on its side with a loud thud. Then the head is twisted to one side and the horns anchored in the ground. One man puts his fingers in the nostrils, another holds the rope to its tied feet, and the third holds its tail. The fourth man takes a small sharp knife and cuts a hole in the tough hide, just above the heart.

He then takes a sword, or a long straight-bladed knife, places it in the hole and with all his weight plunges the sword deep into the animal. As the blade punctures the heart the animal's whole body seemingly comes off the ground and the horns dig and twist deeper into the earth. Within a split second, it comes crashing down with a vibrating thud in a cloud of dust.

The animal lies there panting in pain, its eyes watering, with froth coming out of both nostrils and mouth. The dust settles, its eyes appealing, asking, *'Please explain what is going on?'*

The men move to the fire, where they silently drink their morning Tibetan tea. Then they come back to the animal and take their stations. The sword is pushed back in, this time twisted deeper into the heart. Now there are a less spectacular body jerks, more a series of shudders. Helpless eyes talk directly to you, asking if there has been a misunderstanding. You want to scream, *'Stop!'*

But no, I am but a passing spectator in an ancient practice of the Himalayan peoples. Some villagers choose to push the animal over a cliff - a quick accident. This exposes those doing the pushing to some danger, as the yak, a powerful animal, will be reluctant to follow this more humane alternative to end its life.

I ask, *'Why not shoot it, or cut its throat and be quick about it?'*

'Government no let us have gun. If we cut throat, blood pump onto ground, lose many sausages.' He replied.

The weak heartbeat slowly pushes the sword back out, only to be pushed back in like a hot knife into butter. The animal groans, jerks, tries to raise its head, shit slowly seeping out of its bowels, urine staining the dry dusty earth, heavy breathing, eyes appealing. Again, a look of utter disbelief, *'Why is this happening to me, what is the purpose, and when are you going to stop this foolishness?'*

Then one of the men approaches with a kettle of boiling water and pours it down the animals nostrils, it shudders as its lungs fill with scalding water. Eyes weeping, nostrils foaming, it struggles to raise itself, desperately looking for a last way out. You witness a surge of energy and sense the burning horror of a silent scream. But no one is listening. They drink tea. I stand uncomfortable in my helpless silence. The animal now slumps, growing smaller, air no longer filling its once proud powerful lungs, lying there in the dust, restful, eyes now glazed milky-grey - soul departed.

One of the four recites time immortal prayers over the body, places rancid butter on the horns and on the hooves, casting tsampa powder into the air. An offering to the gods, and there is peace, they are genuinely sincere and all are respectively silent.

They peel off the hide, scrape it, lay it out to dry, cut open the belly, take out all the internal organs and scoop out a couple of buckets of rich red blood. Then they cut up large sections of meat to be hung and cured over smoky yak dung fires, while securing all the fresh steaming manure for the fields. Nothing is wasted.

Slowly, more village people come to collect the pieces of meat ordered the previous evening, before the 'accident' happened. There is a festive feeling in the air - people chatting, smiling and laughing. The four have chosen some delicacies, and set about boiling what looks like an octopus in a giant black pot - Nepali haggis!

Tashi Lama

Chapter 4

Tashi Lama

In the town of Manang I usually stay at Mavis's Kitchen, where she has five small bungalows placed in her vegetable garden. She is short in stature, large in personality, generous by nature, has an infectious laugh and runs the best eating place in town.

One morning before the cock crowed, I walked up the hillside behind Mavis' Kitchen to visit Tashi Lama. Tashi is in his seventies and lives in a series of small caves in the rock face high above the town with his daughter. His wife, sadly, had passed away a few years ago. I climbed the ladder up to his living room, which is a small cave open at the front, and greeted him with, *'Namaste'*.

He bows and replies, *'Tashi Delek.'*

My greeting was in Nepali, his in Tibetan.

Tashi does not remember me.

'Norway - Khukuri.' I say.

'Ahhhhh, Burrah Sahb... - Khukuri...' his face lights up, *'Velkome, yes ys, kom com...'* he jumps up, *'kom com Gorrah Sahb...'*

Thirty years earlier, I had presented Tashi with a short-bladed knife with a stubby wooden handle, which he had laughingly called a 'Norway-Khukuri.' The real khukuri sports an impressive curved steel blade and is the symbol of a Gurkha soldier's bravery.

I had also taught him Gammeldans, which is Norwegian country folk dancing. With my small tape recorder providing music, we danced and danced, while his wife clapped and clapped and laughed until tears of joy ran down her face.

This time his daughter serves Tibetan tea, accompanied by the dreaded words, *'Shay shay, shay shay'* - more more! Tashi ignores my facial grimaces and my Donald Duck sounds as I gulp it down. It would be impolite of him to comment on his visitor's uncouth manners. She smiles, understandingly compensating my suffering with a plate of crispy tasty peas in their pods.

I doze in the morning sun while Tashi dons his robes to greet the first tourists, who we can see toiling up the barren hillside below, on their way to receive his blessings.

Room with a view

Tashi's chanting, the prayer flags flapping in the wind, the Himalayan griffon circling high above and as the taste of the Tibetan tea fades, I burp contentedly.

Sitting in a cave high above the town of Manang
In front - rising from a barren valley floor
The mighty Annapurna range
Undulating rock and ice
A majestic snow covered fortress

Tashi Lama is in his small cave chapel
Chanting time immortal prayers
Buddhist prayer flags flap in the wind
Against an eternal turquoise sky
Five colours
Five pure lights

Blue *for sky*
White *for wind*
Red *for fire*
Green *for water*
Yellow *for earth*

Pabbhassara - *The Essence of Light*
Completion of the Aura spectrum
In, but not of serendipity
We are but specks of mortal time
A crow flies by
A soul against a Himalayan sky...

Chapter 5

Manang - People, land and culture

Sitting in Tashi's cave, before me The Great Barrier rises 3000m from the valley floor to a height of over 7000m above sea level and stretches over 20km from east to west.

Below lies the town of Manang, located in the northern central part of Nepal on the Tibetan border and flanked by the Annapurna massif to the south. To the north rises the rugged Chulu range, peaking at 6584m and behind it is the endless Tibetan plateau.

The Upper Manang or Nyeshang Valley has been sculptured by two rivers and is the home of seven high altitude villages. In winter, life is a struggle, and many families move to Lower Manang with their cattle and goats. Here, the Marsyangdi River flows and this is where we find Chame, the district's administrative centre. Nearby, hidden away behind a giant rock fortress, flows the Nar Khola river.

Along its banks, the trail leads to a once orbidden kingdom and the villages of Nar and Phu.

The area also housed the last of the fighting Tibetan warriors, the Khampas. From this lost valley in Nepal, they continued the unequal fight against the invading Chinese. The Chinese don't mess around, they simply entered the Khampa's encampments in Tibet and executed a large number of men of fighting age. No men, no fighting! A pretty traumatic turn of events for the fighters based in Nepal, and then, in their usual manner, the American CIA withdrew their support, resulting in the Khampas being neutered - and eventually scattered.

Today, their sad, empty fortress villages and barren fields bear witness to a lost cause.

A tragic story - This took place in 2009, and centres around yarchagumba, which is fondly referred to as Himalayan Viagra. This rare medical fungus can, according to Tibetan medicine, cure ailments ranging from tuberculosis, asthma, hepatitis, anaemia, emphysema, HIV/AIDS, hair loss and, of course, it has extraordinary libidinous powers. It is this claim to increase potency that is the main reason that it fetches over US dollar 20000 per kilo, in the end marketplace.

The ghost moth and the fungus, cordyceps sinesis, thrive at altitudes between 3300m and 5500m, in moist soil. What happens is that when the ghost moths lay eggs in the grass, the larvae hatch and then burrow into the earth. In the soil, there are microscopic fungal spores which infect the larvae. These parasitic fungi devour the caterpillars from within, until only an exoskeleton is left. When spring comes, a stalk develops and pushes its way through the mummified caterpillar's head to reach ground level. When the stoma or stroma ripens, spores appear which are picked up by the wind and spread further, to seep back into the ground with the next rain or snow melt and the cycle repeats itself.

The temptation of easy earnings led seven young men from the village of Laprak, in the Gurkha district, into Nar-Phu to collect the yarchagumba. A goat herder saw them and informed the village headman of Nar. The Nar people regard the harvesting as their right, and tradition demands that one man from each household participates in any conflict or negotiation. So, over sixty men rushed in search of the thieves. On seeing the village posse, two of the Laprak men ran until they were forced to stop at the edge of a steep cliff.

Here, they were faced with two options: jump and be killed, or face the posse. They chose the latter, but the end result was the same. The villagers stoned them to death and threw the bodies into the gully.

In the meantime, the other five were taken prisoner and marched to the village. Here, the victory was celebrated with chang (millet beer) and potent rakshi (rice wine). Then the obvious question arose, *'What to do with the captives?'* If freed, they would inform the police and a murder investigation would ensue.

What followed became the darkest chapter in the area's history. All five, between the ages of 15 and 18, in front of each other's eyes, were savagely stoned and hacked to death with knives and then buried.

Later, when the young men were reported missing, it was assumed they had travelled to Kathmandu in search of work and the bright city lights. But a seasonal worker told a friend about the incident and rumour spread, which the police in Chame picked up. A force of over twenty policemen was dispatched and they found the remains of the first two victims. Once the locals were aware that the police were in the district, they dug up the other five bodies, cut them into bits and threw them into the river.

No bodies, no proof!

A further 30 police were flown in by helicopter and the village was surrounded. This was followed by a skirmish, resulting in about 70 men being arrested and then jailed. As the police investigated, the majority were released over a three year period, with the two main culprits receiving twenty year sentences.

The village of Nar ©JH

Lower Manang - Here Nepal's national flower flourishes and graces the country's rhododendron trees. The area abounds with pine and juniper trees, yet even in the lower regions, a tree takes 50 years to grow to maturity. The area is also rich in herbal plants, which play a significant role in the health of the people. Rosa seresa and seabuckthorn are rich in vitamins and others are used for poultices, painkillers and antiseptics. Birds are plentiful, with the magnificent Himalayan griffon seen gliding in the warm air currents way above you.

Indigenous wild animals include the elusive, protected and endangered snow leopard, which preys on domestic goats and young yaks. The locals take the law into their own hands to protect their livestock - with the bonus that the pelt fetches a princely sum. Its claws and teeth being highly valued, for shamanic medical practices and Bon-religious rituals. The shy musk deer is also protected, but again hunted, mainly for the secretion from its reproductive glands, which produces a sensual and expensive perfume. The galloping blue sheep are not sheep, but mountain goats, and are relatively common, yet shy. The proud silky brown-coated tahr, on the other hand, is rarely seen, with wolves as well as black and brown bears being even rarer.

We of Shang - The Nyeshangte people of Manang are said to have originated from the Kingdom of Shangshung in ancient Tibet. Their name means 'We of Shang', and many still speak a Tibeto-Burman language. They brought with them the shamanistic faith of Bon, an animistic spiritual religion based on a belief that the universe is filled with good and evil spirits. In this nether world, man walks the earth and the spirits reside in the wind, water, stones and plants.

Later, Siddhartha Gautama (approx. 563-483 BC) was born far to the south in what is today's Lumbini, near the border with India. He was a prince who gave up a kingdom to commit himself to following the path to enlightenment. Siddhartha was never a god, but became the enlightened one: Lord Buddha. One to follow and to emulate and whose teachings reached Manang in the early 10th century. Here, by integrating with the established ancient religion of Bon, Buddhism became the dominant faith.

In a similar evolutionary process, Hinduism adopted the Buddha into the pantheon as the seventh avatar of Vishnu, one of three major gods alongside Brahma and Shiva.

Hotel kitchen. Tshering Dolma was my host in the hotel without the sign.

Today Manang is no longer the medieval fortress town it was when I first visited here over 30 years ago. Then, there had been only two hotels; one with a sign, the other without.

Today, Manang can be reached by road, has electricity, shops and cafes, hotels and restaurants, discos and cinemas, the Internet, modern bakeries and a nearby airport. Some feel that all was better before and should remain so. One can feel and express concern when traditional values are continually being challenged and eroded. Yet there will be change, with or without tourism and we who travel here are all tourists.

Nyeshangte women - They have a central position in Manangi society. The girl may freely decline an offer of marriage brokered and agreed between parents. If she does consent, then the couple may live with the girl's parents before they are formally married. There is no dowry in Nyeshang culture and both men and women bring property into the marriage. Many Manangi men are pretty wild and are usually associated with drinking, gambling, hunting and war, whereas the women are regarded as better Buddhists because they practice the arts, song and dance, plus they follow the religious rituals.

Attendees of Badhe Festival, 1979. Photo: Zdeněk Thoma.

Religious festivals play a key role and the recently revived Badhe Festival is based on a fight between two brothers, telling how the conflict was resolved by the wise ruler and neighbours. The theme is that the collective wellbeing is of greater importance than that of the individuals. When taking into account the harsh winters and the hard lives the Nyeshang have led over the last 500 years, then it is easy to understand the binding importance of such festivals.

Mavis' Kitchen - One evening sitting with Mavis, we talked about Maurice Herzog and the French expedition that had first climbed Annapurna 60 years earlier. Most local people and tourist trekkers are unaware of this historic ascent. We played with the idea of my writing a small booklet about the mountain and its history. I would follow Herzog's route in reverse, across the Tilicho Pass, and then proceed to the original French Base Camp on the other side. Once written, Mavis could sell the booklet to tourists and local people alike.

The next few days were spent fattening myself up on Mavis' food and drinking countless hot lemons laced with honey. The last thing was to buy the all important Mars bars.

The Paradox of Our Age

We have bigger houses, but smaller families;
More convenience, but less time.
We have more degrees, but less sense;
More knowledge, but less judgment;
More experts, but more problems;
More medicines, but less healthiness.
We have been all the way to the moon and back,
But have trouble crossing the street to meet
the new neighbour.
We have built more computers to hold more
information to produce more copies than ever,
But have less communication.
We have become long on quantity, but short on quality.
These are times of fast foods, but slow digestion;
Tall man but short character;
Steep profits, but shallow relationships.
It`s a time when there is much in a window,
but nothing in the room.

His Holiness the 14th Dalai Lama

ANNAPURNA

'Regards Vers L'annapurna' by Maurice Herzog and Marcel Ichac.

Dhaulagiri, 8167m. The French first tried Dhaulagiri, declaring it to be 'unclimbable', then they turned their attention to Annapurna.

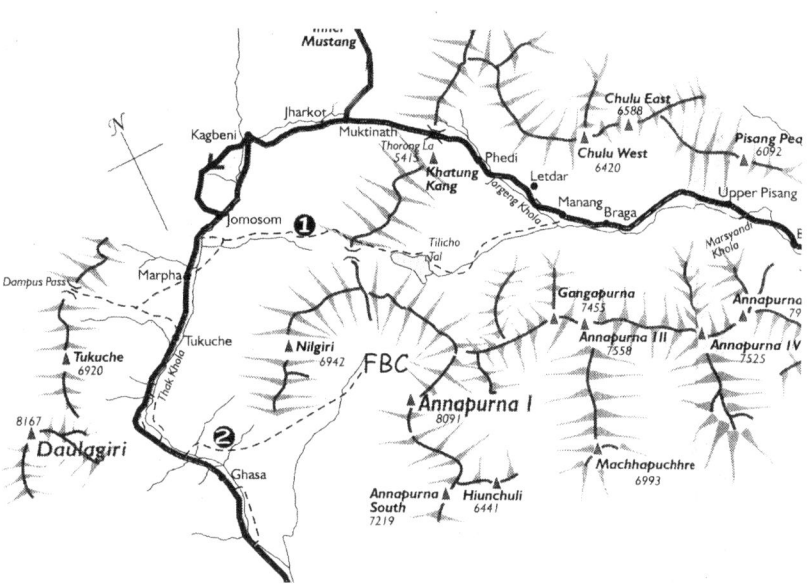

Annapurna *(8091m) is the highest peak in a complex massif consisting of 13 peaks over 7000m and 16 over 6000m, plus numerous vast glaciers and ridge systems, deep valleys and raging rivers.*

Key: *Todays trekking route around Annapurna.*
❶ *Route I took over Tilicho, and* ❷, *followed to the French Base Camp - FBC*

Chapter 1

Annapurna - In the footsteps of the French

I waved goodbye to Mavis and headed off to the village of Khangsar. The French expedition had reached this village after having crossed from the opposite direction, via Tilicho Lake, in an attempt to find Annapurna itself. At that time, the village people had never seen a white man, so the sudden arrival of the French from over the high snow covered passes caused quite a stir. Rather similar to a two headed green spaceman landing in your town square to buy groceries. Herzog did not have two heads, but he and his expedition were low on food, and they were looking for supplies. Unfortunately, the villagers were so poor they had virtually nothing to sell.

I, on the other hand, ate well, slept soundly on a foam mattress in a lodge that had a toilet, shower and electricity - and woke to the smell of Nescafe. What change sixty years has brought.

In my rucksack was a small tent, weighing 2kg, a foam mat, a sleeping bag and a water bottle, some food, clothes, crampons, an ice axe and two books. The total weight was about 10 to 12kg. Cooking my own food would mean having the extra weight of a stove, fuel and pots, plus my having to wash them. So my plan was to eat at the next two lodges and then buy a week's supply of momos, a traditional Tibetan dish, which I would eat cold.

The next lodge was filled with a group of tourists who had ascended too high too quickly. The majority seemed to be suffering from a degree of altitude sickness, as the room was thick with pungent smells. An obvious symptom of H.A.F. - or High Altitude Farts. If I was not to die of gas poisoning, it was better to continue after eating some soup and Tibetan bread. I camped further down the trail, where the silence of the night was disturbed only by the occasional avalanche thundering down the distant slope. Sleep just came.

The trail now meandered its way across a lunar-like landscape and led to the so called Tilicho Base Camp, and the last lodge before Tilicho Lake. Here, there were two lodges, one for tourists and one for Nepali porters. I pitched my tent by the latter, which was run by two charming sisters.

Most trekkers take the steep morning hike up to the lake, but do not cross the Tilicho Pass, owing to the lack of lodges. It's a pretty desolate place and an unexpected snowstorm could and has resulted in death and serious injuries. The sisters fried forty momos, boiled twenty small potatoes and four hard boiled eggs for me. The next morning, I took the 3 hour walk up to the ridge overlooking the lake. One of the sisters offered to carry my sack halfway for a few dollars. I accepted, on the understanding she promised not to tell anyone.

'Alps upon Alps, peaks upon peaks' - My tent was a solitary silhouette set against a white background of steep snow slopes that rose out of a sea of ice. Inside I settled down to find a packet of coconut biscuits in my rucksack. What a pleasant surprise!

The warm morning light brought with it weird sounds. Was it the ice monsters talking from the depths of the frozen lake? No, it was the recent day's rising temperatures causing the ice to stretch and groan. Breakfast consisted of two chapattis and a hardboiled egg, while the day was spent reading and acclimatising. Early the next morning, I walked 4km along the northern shore towards the western end, here to be stopped by an impressive vertical 200m high rock cliff rising out of the lake's waters. This was obviously the same cliff that had forced Herzog out onto the ice, coming the other way, sixty years earlier.

Today's goal was to reach the other side. I am by nature a competitive person (with myself), so I tentatively went out onto the ice and moved towards the centre of the lake. Ten very long minutes later the reality of my insanity began to sink in, and a voice spoke from above:

'Hey stupid! Why are you in the middle of a melting ice lake high up in the Himalaya, all on your own?'

Try to visualise, as I found myself doing, a small, and rather pathetic, lone figure going through the ice: screaming, arms waving, desperately trying to get back on the ice, slipping back, a cry for help, looking around, feeling stupid, legs going numb, becoming weaker, getting angry, desperately clutching. What will people say? Acceptance, frozen fingers, a tear, a deep breath, farewell Therese, farewell Filip, aware of what I was about to lose, before slowly slipping under the water, into the darkness, no longer cold, a helpless white face, bubbles …release …idiot …no more.

Reaching the safety of the other shore, I ate a Mars bar and, with my sugar level restored, moved southwards towards the Tilicho Base Camp. I turned east, put on crampons and ascended the glacier that formed the southern shore of the lake.

Crossing a glacier alone may sound irresponsible, but on a dry or naked glacier, one not covered by snow, then all the crevasses are visible. So if you fall into a crevasse it is your own (stupid) fault. Five hours later, as the evening light crept across the sky, I snuggled into my sleeping bag to eat ten Tibetan momos and a can of sardines, with a Mars bar for dessert.

Tilicho Lake, with Mt. Tilicho, 7140m, in the background. Part of The Great Barrier that confronted the French during their search for Annapurna.

After crossing the frozen lake, I followed the edge of the glacier - keeping out of avalanche danger. A spooky, yet enjoyable outing.

I was up long before the sun and packed my gear, intending to repeat yesterday's lake crossing before the sun warmed the ice. It soon became obvious that the last week's warm weather had melted the ice at the edge to a degree that to step onto the ice would be a risky exercise. I swallowed my pride and backtracked, reaching the small trail that led towards the north pass two hours later. I pitched the tent, ate a light lunch and slept.

The journey continued before the sun rose, up a steep path that wound its way through a desert-like landscape, now clearly marked by boot prints. I slowly worked my way up the steep and loose hillside, huffing and puffing. At the first pass, the weather began to change, the path was visible but disappeared under the snow half an hour later. The mist was getting thicker and I plodded on, frustratingly breaking through the snow crust, up to my knees, time and time again.

When the path re-appeared there where two routes to choose from. One led to the right and the north pass. The other led straight on, west, towards the Mesokanto Pass. This was the pass Herzog must have crossed 60 years ago, almost to the week. An obvious choice.

The wind died and the land was soon totally engulfed in the afternoon mist, making it difficult to navigate the featureless landscape. To retrace would mean a night out at over 5000m, something to avoid, as there was the smell of snow in the air. Even in the all-encompassing mist, the non-marked way felt right, so I pushed on. Then, reassuringly, the path reappeared as it crossed a series of rough sandy gravel slopes.

Time to rest and time to ask, *'What greedy bugger has eaten my last Mars bar?'*

Before reaching the Mesokanto Pass the mists thinned out to ghostly Banshee whispers and then disappeared completely - the late afternoon light was soothingly kind. On reaching the pass, I was tired and feeling a little chilly. So it was a relief to be greeted by a spectacular 700m snow slope that formed an ideal toboggan run down the other side. This was an unexpected bonus. The wind was building up, let's get out of here! With the rucksack held tightly to my chest, I dropped onto the seat of my pants and shot down the snow slope seemingly at the speed of light - a magnificent toboggan run - wheeeeeeeeeeeeeeeehiiiiiiiiiiii!

It had taken a whole day to reach the pass, but seemingly only minutes to shoot down the other side. On an island of grass and moss in the midst of a grey lunar scree slope, I pitched my tent. The air was Mediterranean warm. I ate what was left of the momos and dozed off with the tent door open. Waking early the next morning, I was treated to the pleasant sight of a grazing herd of blue sheep, total silence. Ten minutes passed. On noticing the tent, the does suddenly looked apprehensively towards the big buck. He, a superior creature with gigantic curled horns, cast an inquiring glance my way, asking, *'Who are you, what do you want?'*

He rightly sensed that I was not after his harem, so he tossed his proud head and dismissed me as an irrelevant irritation. The herd walked slowly away, ladies first, secure under his watchful eye.

Next stop, the town of Jomsom.

Chapter 2

Annapurna - 1st ascent

Maurice Herzog's book, *Annapurna,* has inspired hundreds of thousands over the years - a classic in every respect.

It was June 1950 when the French team travelled overland from India with permission to climb Dhaulagiri (8167m) or Annapurna (8091m). Maurice Herzog was the leader, and the team included Louis Lachenal, Lionel Terray, Marcel Ichac and Gaston Rébuffat - some of France's, no, some of the world's, leading alpinists.

The Himalaya extends over 2200km and, according to Eberhard Jurgalsky, boasts 241 summits over 7000m. It is generally accepted that there are fourteen main peaks that rise to over 8000m (there are more). By 1950, there had been more than twenty expeditions from various nationalities who had attempted to climb an 8000m peak.

Incredible as it may seem today, the French knew little more than the names, heights and approximate locations of the chosen peaks. Dhaulagiri, *The White Mountain,* was their first choice. One can imagine their reaction on first sighting, as it towers 5000m above the main valley floor. They made numerous attempts to find a way on to the mountain itself, until they concluded that Dhaulagiri would 'never be climbed!'

Their attention then turned to Annapurna, *The Goddess of the Harvests,* or *The Mother who Feeds.* Like most of the Himalaya, the area was virtually unexplored and unmapped. Yet the map of the day did show Annapurna overlooking the Tilicho Pass, from where the team assumed they could gain easy access and start the actual climbing. They hired a local guide to make an exploratory probe and he soon got them lost. They were confronted by a giant maze of walls, valleys and moraines. Eventually, using trial and error, they reached a height just below 5000m. Here they were met by a 4km long and 1km wide frozen expanse of water - a lake, which was not on the map!

Their second surprise, also not on the map, was a barrier, an extension of the Nilgiri Mountains. This was 2000m high rock and ice wall, which rose out of the non-existent lake to heights of 7000m and continued as far as the eye could see.

The third surprise - There was no sign of Annapurna herself. Even so, they decided to push on and try to find a route through the barrier on the other side of the lake, with the hope of reaching the elusive Annapurna. To their right, avalanches swept down onto the glacier and to their left was a 200m high vertical rock wall growing out of the lake. The Sherpas, while brave and courageous on snow and ice, must have been horrified on realising that Herzog intended to cross an expanse of frozen water - and to take them with him!

The French and their Sherpas crossing Tilicho Lake, 4940m, in 1950. I crossed at virtually the same point, 60 years later.

With the lake behind them, they pushed on, in the hope of finding a break in what we now call The Great Barrier. Alas, to no avail. At this point, after many disappointments, the majority of us would have given up and gone home. But not the French. They re-grouped, returned to their base camp, reorganised and started a new approach under the southern flanks of Nilgiri. There were no trails and the terrain was complex, steep and loose. Finding campsites with water was a challenge. Eventually, they were confronted by a frightening labyrinth of rock and ice, with avalanches raging down the slopes.

Roc Noir. *Part of The Great Barrier, also not on the map.*

Again, lesser men might say, *'Ok, been here, seen it, don't like what I see, bad idea. Let's go home to the wife and kids.'*

They did not, they simply split the party. Some probed the mountain's defences, while the others moved the equipment up from the village of Tukuche. They established base camp and, in rapid succession, camps 2, 3 and 4. Both Sherpas and French climbers worked together - breaking trail, establishing camps up the treacherous slopes, sharing the risks, the frustrations and the hardships.

At 7000m, under camp IV. A Sherpa is stopped by a layer of ice under the snow. Despite the wind and cold Gaston Rébuffat captured this classic image of climbing on Annapurna.

The Summit - From Camp 5, after a sleepless night, with the hell of the monsoon about to break, Herzog and Louis Lachenal headed for the summit at 6am. Worried by the sensation of approaching frostbite as they ascended the enormous expanse of the North-East Face, Lachenal asked Herzog what he would do if he turned back.

'I would go alone.' Herzog replied.

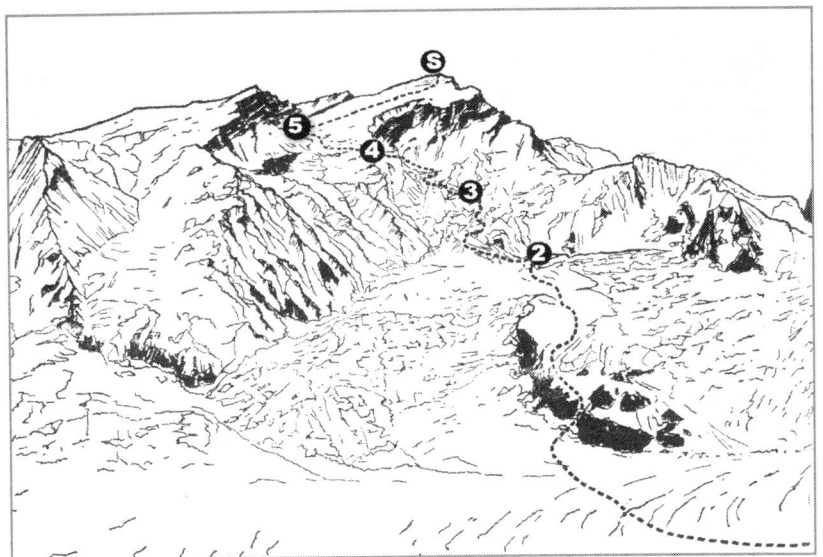

Route followed by the expedition with camps 2, 3, 4 and 5 shown.

Herzog later wrote:

'A false move would have been fatal …where was the top? …leaning on our axes, we try to recover our breath and to calm down our hearts, which were thumping as though they would burst …a slight detour, a few more steps …the summit ridge came gradually nearer …we dragged ourselves up …could we possibly be there?

'Yes! A fierce and savage wind tore at us. We were on top of Annapurna… Our hearts overflowed with unspeakable happiness… The summit was a corniced crest of ice, and the precipices on the far side, which plunged vertically down beneath us, were terrifying, unfathomable. Clouds floated half-way down… Above us there was nothing.'

The first ever 8000m peak had been climbed.

The way down - It was 2pm and things began to go wrong. Herzog lost his gloves, the weather changed and soon they were enveloped in a thick, swirling cloud and freezing winds. On reaching Camp 5, Herzog found that his hands were, *'…violet and white and as hard as wood.'* Luckily Rébuffat and Terray had come up to Camp 5. That night a violent storm raged around their tent.

Reinhold Messner in his book, *Annapurna - 50 years of Expeditions in the Death Zone*, notes, *'… without oxygen masks, their descent was reduced by frostbite, falls, new snow, avalanches, fog and snow blindness to a crazed stumble through the Death Zone.'*

The next day, they descend through the steep ice and snow, with avalanches thundering all around them. With a second night approaching, they begun to dig a snow cave when Lachenal fell through a crust of snow and into a crevasse. Luckily, it was only 5m deep and offered protection from the elements. All four stuffed their feet into the one sleeping bag they had.

The next day, Herzog, *'Lachenal too was affected by snow blindness... Rébuffat (snow blind) went ahead by guess work, with agony in his face, but he kept on... We had passed the danger zone... The sun was at its height, the weather brilliant and the colours magnificent. Never had the mountains appeared as majestic as at this moment of extreme danger.'*

Avalanche - They thought they had reached safety, when just above Camp 3 an avalanche swept Herzog and two Sherpas 150m down the face.

'All at once a crack appeared in the snow under the feet of the Sherpas, and grew longer and wider. Then I was lifted up by a superhuman force, and, as Sherpas disappeared before my eyes, I went head over heels. Could not see what was happening. My head hit the ice... I could no longer breathe... I turned round and round like a puppet. In a flash, I saw a blinding light of the sun through the snow which was pouring past my eyes. The rope joining me to Sarki and Aila curled round my neck - the Sherpas shooting down the slope beneath would shortly strangle me, and the pain was unbearable. Again and again I crashed into solid ice as I went hurtling from one serac to another, all the snow crushed me down. The rope tightened around my neck and brought me to a stop. Before I had recovered my wits I began to pass water, violently and uncontrollably...'

From base camp, they started the retreat, crossing raging rivers, with the always present danger of landslides. They had to carry Herzog and Lachenal on their backs and on homemade stretchers as they cut paths through steep mountain vegetation. The monsoon had broken, but torrential rains did not dampen the smell of rotting frostbitten flesh. We can almost feel the pain of the bit by bit amputations during the long evacuation.

Herzog lost toes and fingers and Lachenal toes.

BARBIE GOES TO THE HIMALAYA

Annapurna Base Camp - Korean Expedition. Photo: D Durkan

Chapter 4

The Road to Annapurna Base Camp

Walking down the main street of Jomsom, dirty, tired and sweaty, late in the day after thirty days trekking on my own, with the wind blowing in my face, lips cracked, I almost expect to meet Butch Cassidy and Sundance Kid, revolvers blazing.

Jomsom is now well paved, has an airport, hotels, barbers, tailors, Internet cafes, bakeries and a bus stop. My first priority was a cold Everest beer and then to phone Temba in Kathmandu, *'Good if you could meet me in Jomsom in three days. Bring a stove, paraffin, cooking gear and what you need for yourself.'*

As usual, Temba turned up on the dot. We bought porridge, rice and dal (lentils) for ten days, hired a local porter and headed off. We spent five wonderful days working our way towards the original French Base Camp.

The French, noted 60 years ago, that the *'terrain was complex, steep and loose'*. An understatement and they had had nothing to guide them. Complex, yes, just one nondescript ridge followed another, crossing steep loose terrain with a raging river far below, and steep walls all around.

Klondike village - On arrival, we were confronted with a Klondike village of expedition tents. Here Temba met up with friends from his district and we were soon tucking into international expedition cuisine: Norwegian salmon, Japanese sushi, Ukrainian goulash, Russian caviar and lashings of vodka. There was espresso coffee machines, electric lighting, hot solar showers and the Internet.

There were about seven expeditions, of which the Korean Expedition was dominant. They had established a TV studio with full production facilities and a main street paved with wooden-slat boards. The studio was staffed by 20 Korean TV personnel and another 20 staff at two sub-studios at slightly higher altitudes.

The French had no trails to follow and the 'terrain was complex, steep and loose'. We had a faint trail, yet the approach is more demanding than the approaches to the other 8000m base camps in Nepal (except Dhaulagiri).

From B.C., we took a day trip to Camp 1 for a view of Annapurna.

Oh Eun-sun claimed to have already climbed thirteen of the fourteen 8000m peaks. To conquer the remaining one would make her the first woman in the world to have stood on top of all fourteen.

The Korean team had two stated objectives:
1. To get Oh Eun-sun to the top of the mountain.
2. To transmit her 'historic' ascent live, on Korean national TV.

Korean and Nepali National TV studio - transmitting 'live'.

The next day, Temba and I took a day trip, using the expedition's pre-fixed ropes. We ascended the rocky slabs and walls, crossed the dark grey glacier, and passed tottering ice towers, to reach Camp 1. What a picnic spot, what a magnificent mountain - what a sight.

We ate our lunch pack in the middle of the glacier before we descended to a barbeque dinner at one of the expedition camps. Here we learnt that I was probably the only westerner who had actually walked into base camp. Seemingly all the members of the various expeditions had been flown in by helicopter. On landing they went to the high altitude research station to take various oxygen saturation blood tests before taking their doses of acclimatisation medicine. In the meantime, the Sherpas had been fixing ropes and establishing camps up the mountain.

Two days later, Oh Eun-sun was going for her summit push, and excitement in the TV studio was rising. We could hear her breathing and coughing 3000m above us. Producers, directors and technicians were rushing around like headless chickens. Via the TV monitors, we watched Oh Eun-sun crouched in a shelter dug out by the Sherpas, to protect her from the wind. They had the latest weather forecasts, so they knew exactly when to expect a window of better weather at the summit.

'You don't need a weatherman
    to know which way the wind blows.'   - Bob Dylan

Half an hour before the predicted lull in the wind, we watched one of the Sherpas casually coming down from the summit. He had been to the top to check the ropes they had fixed the day before. We saw him carefully cover the ropes with snow, to hide them.

Then, as forecasted, the wind died down and Oh Eun-sun started to move, like a giant panda. Twenty minutes later, we watched her lose her balance and stagger off the customised footsteps. With a Sherpa's help, she regained the pathway. Another Sherpa clipped her out of one rope and into the next fixed rope, adjusting her harness and oxygen set. When prepared, she moved again, only to get entangled in the two ropes. We could see her irritation and sense her fear.

On five monitor screens, we viewed what was being transmitted 'live' to the people of South Korea. The section where she had staggered off the path had been edited out. What was shown was her forcing a new path through virgin snow as she bravely leads her team of Sherpas towards a glorious victory for the greater glory of Korea!

What a market - Mountains cover 70% of South Korea, and the country has a population of nearly 90 million. Mountain hiking is one of the most popular forms of exercise, with approx. 14% of the population being active hikers. The Korean Mountaineering Support Centre notes that mountain climbing has a high status, 43.5% men and 35.6% women look up to the sport. Outdoor sports gear is standard street wear.

The nation's outdoor industry has a yearly turnover in excess of 3 trillion Won (US$2.6 billion), offering potentially great economic gains for a megastar woman mountaineer. We have nothing like this phenomenon in the West, but it could be compared to the magnitude of Madonna going mainstream market. With a successful ascent of Annapurna, Oh Eun-sun would cement her position, and became Korea's version of Evita.

However, her earlier claim to have climbed Kanchenjunga was being challenged. Based on the height she was last sighted at, the time taken to reach the summit and then return was all felt to be too fast. Norwegian Jon Gangdal, who used her expedition's fixed ropes a week later, observed her 'summit' pennant 50m below the actual summit. She had no photographic evidence, and one of her Sherpas stated that she had not summited.

She claimed the density of the mist had made it difficult to orientate, and that the Sherpas had 'informed her' that her high point was the summit. Ok, but if she had been in doubt she could have walked a few more steps to see if the slope started going down the other side. A possible sign that she had reached the top. Alas no, her 'summit pennant' indicated that the summit was elsewhere than where she had stopped.

In all likelihood, she was totally exhausted, could not continue, and the Sherpas wisely decided to shepherd her down to safety before she collapsed. The Korean Alpine Club declared her ascent void. Therefore, Edurne Pasaban became the first woman to have climbed all 14 of the 8000m peaks, in 2010. Of more worthy note is Gerlinde Kaltenbrunner, who in 2011 completed her last of all the 8-thousanders, all without the use of supplementary oxygen - an amazing feat.

Candle in the wind - Life at Annapurna Base Camp was great. We played poker, volleyball and watched CNN. Then the news reached us that a climber was trapped below the summit and conditions were too extreme to reach him. Soon the base camp fraternity were on the Internet telling friends, family, sponsors, the media and the world that someone was dying on the killer mountain. The mountain was too dangerous this year, so they may have to abandon their expedition.

The weather broke, there was nothing one could do - his life was like Elton John's *Candle in the Wind*, snuffed out.

It was time to go home.

The style and the spirit in which the mountain is approached and how it is treated has great intrinsic value, which we should acknowledge, honour and strive to protect.

Book 4

Instant Mountaineers
'Just add Gore-Tex'

A mountaineer visiting our land often walks in a bubble
I ask you to step out of the bubble
See, hear, smell and feel the land
Be curious
Don't just rush through
Stay with us, live with us
The high peaks have their own laws
They do not care if you climb them
It is only you who think that what you do
has any value or importance
Learn to know the mountain, learn their every humour
Experience their strengths, their might, and honour them

Meet the Parbat - meet yourself

Swami Kailash

Chapter 1

Instant Mountaineers

There is a growing trend, across all nations, by individuals and groups in both trekking and in mountaineering, that:

'I have money, I have the latest in gear, I have been on a climbing course and/or been guided up Mt. XYZ and I have a jacket covered with logos. My family, friends and sponsors know that I am on an 'expedition' and therefore it is my right to succeed.'

Seemingly, getting to the top is more important than how one actually reaches the top. And that 'image' is more important than 'content'.

To address this, I lean on Chris Bonington, who in his *Foreword* to the first English edition of *Annapurna* by Maurice Herzog, notes some of the developments which have contributed to the base value(s) we find in Himalayan mountaineering. Those of: exploration, knowledge, companionship, trust and respect, competence, commitment and dedication. Characteristics spiced with the need for ability to judge potential risks, and adventures that are garnished with a little humility.

His words are not exhaustive, as that would require a library, but they do give an indication as to the development that has gone on in our sport. Indicating how this has laid the foundations for those who practice the sport today, in a pro-active and progressive manner.

Reinhold Messner expresses his concern as to the death of adventure in mountaineering - whereby the use of modern equipment has dragged the mountain down to the level of the performer, and not the reverse. He invites you to join him in pure thought climbing.

I defer to these notables - and allow you to absorb the basic values - ones worth the trouble to understand, practice, honour and to protect.

Chapter 2

Foreword by Chris Bonington to the English edition of 'Annapurna' - by Maurice Herzog

Maurice Herzog's 'Annapurna' is as fresh and exciting to read today as it was in 1952 when it was first published in Britain. It marked one of the greatest breakthroughs in the history of mountaineering, for Annapurna was the first peak over that magic height of 8000m to be climbed. Before the Second World War, greater heights had been attained on Everest and K2, but it's the reaching of a summit that captures the imagination of the public, and in all the attempts on 8000m peaks before the War, each expedition had been defeated, though some had come very close to success.

But Maurice Herzog's book is memorable for much more than the scale of the achievement. It captures the very essence of the adventure. There is a Gallic élan about the team, only one of whom had been to the Himalayas before. They approached the giants of that range as if they had been peaks of the Mont Blanc massif, with the same enthusiastic naïveté as was seen in the very first Himalayan pioneers of the late nineteenth century. They were undoubtedly going into the unknown, for they were the first mountaineering expedition to be allowed into central Nepal. Their maps were inaccurate and they had the greatest difficulty even in finding Annapurna. Once on the mountain, they found themselves in a race against the monsoon and only just snatched the summit in time. They were then caught in an epic descent that led to Herzog losing most of his fingers and toes through frostbite. It is this combination of glorious success and near-disaster that makes the story such a memorable one. But it is the vivid quality of Herzog's writing that makes this book such a great Himalayan classic.

To appreciate fully what the French expedition achieved, it is worth setting it in the context of the development of Himalayan climbing prior to the start of the Second World War. The first serious attempt on an 8000m peak was made in 1895 by the famous British alpinist, A.F. Mummery. Strangely enough, his attempt demonstrates the way Himalayan climbing has gone full circle, for the drive in climbing development today is towards an alpine style of climbing on the highest mountains. In 1895, however, they were not ready for it.

There were just three in the team, together with two Gurkha porters, and they were attempting Nanga Parbat, at that time the most accessible of all the 8000m peaks. They first looked at the huge Rupal Face of the mountain, but dismissed it as too difficult, and subsequently crossed the Mazeno Pass to reach the Diamir Glacier. The team tackled a rock spur on the face which, over eighty years later, the famous Austrian climber, Reinhold Messner, was to descend after making the first ascent of the Rupal Face. Mummery, however, was discovering from direct experience the problems of altitude and the colossal scale of the Himalayas. He retreated, but didn't give up and, with the two Gurkha porters, crossed the Diamir Col to explore the northern side. This was the last time he was ever seen and the cause of his death is as great a mystery as that of Mallory and Irvine on Everest in 1924.

The next attempt on a major peak was in 1902 with another small British expedition led by Oscar Eckkenstein. This time their target was K2, the second highest mountain in the world. Once again, they didn't really know what they were taking on, and it is not surprising that they only reached a height of around 6600m. An interesting aside on the expedition is that, one of the members was Alesteir Crowley, who achieved considerable notoriety as a Satanist. On this expedition, he threatened one of the members with a loaded pistol when the unfortunate man talked of retreat.

The two world wars marked very definite stages in the development of mountaineering, as of so many other aspects of human evolution and history. The post First World War expeditions saw the beginning of the use of systematic siege tactics on Himalayan peaks, started on the north side of Everest by British expeditions and pursued, particularly by the Germans, on Kangchenjunga and Nanga Parbat.

The tactics involved the setting up of a series of camps, the use of fixed ropes in places, and the development of a corps of specialist high-altitude porters. On Everest, oxygen was used for the first time in 1922. There were, in all, seven British expeditions to Everest between the two wars. Their approach was through Tibet, since Nepal was closed to foreigners. On first reconnaissance, they discovered that the most obvious route was by the North Col, and all subsequent expeditions attempted this route. They were beaten by a combination of altitude, snow conditions, and the rock structure of the upper part of the North East Ridge. Of these seven British expeditions, three reached a height of over 8000m. In 1924, Mallory and Irvine, using oxygen, achieved at least 8,600m, when Noel Odell last saw them.

It seems unlikely that they made it to the top, but this can never be totally discounted. On the same expedition Norton, without oxygen, climbed to 8350m, and in 1933, three climbers reached around the same height. But there seemed to be an invisible barrier stopping any of these pre-War expeditions getting all the way to the top. It was probably a combination of factors: the poor equipment of the time, bad luck with the weather and snow conditions, and the fact that not enough was known about high altitude climbing in this period.

Everest was very much a British preserve, and so climbers of other countries had to look elsewhere. Only three 8000m peaks were attempted: K2, Kangchenjunga, the third highest mountain in the world, and Nanga Parbat. German and Austrian expeditors were the most active, which is not surprising, since during this period climbers from these two countries were making technical advances well ahead of those of almost any other nationality, except perhaps the Italians.

There were two German expeditions to Kangchenjunga, reaching a height of 7700m on the North-East Spur, and three expeditions to Nanga Parbat. On the 1934 expeditions to the latter, led by Willi Merkl, he, Welzenbach and Wieland reached the base of the summit pyramid at 7848m. They were only 250m below the top, but dropped back to where the other two Germans and eleven Sherpas had pitched the top camp. The following day a fierce storm broke, and in the desperate struggle to retreat over the subsequent days, Merkl, Weiland, Welzenbach and six Sherpas died from cold and exhaustion. It was the worst ever accident in the Himalayas, and emphasised the hidden dangers of these technically easy, but very long and complex routes from which, it is almost impossible to retreat during severe storms. Herzog's team were to have a similar experience in 1950.

Perhaps the boldest, and certainly the most elegant, pre-Second World War expeditions were on K2. Since Eckenstein's first visit, the Italians had made three reconnaissance expeditions up the Baltoro, but, in 1938, it was the Americans who made the first serious attempt, with a compact expedition of only five climbers led by Charles Houston, accompanied by six Sherpa high-altitude porters. Houston had been on Nanda Devi in 1936, when Bill Tillman and Noel Odell reached its 7816m summit. It was the highest peak to be climbed before the War, and in many ways, the attempt was in the idiom of the modern compact expedition.

Houston, unaffected by the traditions and state of mind that inevitably influenced expeditioners from the Alpine based countries, was inspired to tackle K2 by Bill Tillman's approach to climbing.

His team was very nearly successful as their high point was 7924m, just near the foot of the summit pyramid, on a climb that was, undoubtedly, the most difficult technically that had been achieved in the Himalayas' up to that date.

A second American assault was made in 1939, by a completely different expedition led by the German-born Fritz Weissner. He was probably the most talented of all the pre-War mountaineers, having pioneered extremely hard rock climbs in Saxony, and the Eastern Alps, and having been a member of the 1932 Nanga Parbat Expedition.

Unfortunately, however, his team members were comparatively inexperienced. In spite of this, climbing with his Sherpas he pushed the route up to the base of the pyramid and very nearly managed to reach the reasonably easy ground that would have taken him to the summit, but the climbing had proved too difficult for his Sherpa companion, Pasang Kikuli, who insisted on retreating. The only other climber on the mountain at this stage was Dudley Wolfe, a rich American who had, in part, funded the expedition. He was one of the least experienced members of the team and probably should never have been allowed to get up to the penultimate camp (VII). When Weissner and Kikuli staggered down to his camp, frostbitten and exhausted, he elected to stay there by himself to await their return for another attempt. It was a lunatic and disastrous decision, since, unknown to Weissner, Jack Durrance, another member of the team, had ordered the Sherpas to strip all the camps below Camp VI.

Wiessner and Kikuli only just managed to reach base camp. They sent Sherpas back up to bring Wolfe down but by this time he was too weak to move. The weather now broke and Wolfe and three of the Sherpas, including Pasang Kikuli, who had returned to the mountain in this desperate rescue bid, lost their lives. Inevitably there were recriminations with Durance, who had made a series of very controversial decisions, blaming Fritz Weissner for the disaster. It is only in comparatively recent times that Weissner has been fully exonerated.

This was the last pre-War expedition to the 8000m peaks. It demonstrated, as did several other of these expeditions, just how dangerous Himalayan climbing can be. It certainly has become no safer over the years. In my own experience of 8000m peaks, someone has been killed on every single expedition. It's a frightening toll which throws into perspective the achievement of all the members of the

French Annapurna expedition who, after Maurice Herzog and Louis Lachenal reached the summit, managed to get them back down despite their state of exhaustion and a raging storm.

Climbing on the Himalayan giants is like walking a shaking tight rope with the potential of disaster constantly with you. So why do it? For the allure of danger, the mystery of the unknown, the sheer beauty of the mountains around you, and the drive of ego. All of this is captured by Maurice Herzog in 'Annapurna'. It is what makes the book one of the great classics of expedition literature.

Chris Bonington

3rd June 1950, Louis Lachenal and Maurice Herzog reached the summit of Annapurna.

But they paid a price.

Chapter 3

'50 Years of Expeditions in the Death Zone'

Reinhold Messner both questions and answers:

'What changes have there been in the field of 8000m mountaineering over the last fifteen years? Equipment has become even lighter, the collective knowledge about the Death Zone has grown... So the major change is that many different groups of people, commercial expeditions, as well as national teams, now lay siege to the highest mountains and leave a lasting impression on them.

Following the first ascent of Annapurna, for the next thirty years mountaineering was almost exclusively reserved for the elite of the mountaineering nations. Who, in the main, attempted to climb higher or more difficult peaks, or pioneer new routes on peaks already ascended.

It was Mount Everest and the so-called 'easy eight-thousanders' that were to attract the crowds, the result of all this attention was that the climbs ended up being pre-prepared both by, and for, these masses of people, becoming relatively easy to find and readily accessible, to the extent that less talented climbers began to arrive on the scene.

...the single file lines of hoards that have reduced the eight-thousanders to prestige peaks for the masses.'

Messner's prediction about mass 8000m peak tourism is now a reality.

Another question he raised, *'Are today's big expeditions manned by the elite of their climbing club or by the mountaineering elite of a nation?'*

No, they are not. The serous high altitude mountaineer is a part of a healthy development, whose main aim is to climb non pre-prepared routes or new routes on peaks that have been climbed previously, or ascend virgin peaks.

No matter which, they choose to climb in a purer style, like using less, or no guide/porter support, Alpine-style, climbing faster, or solo, without supplementary oxygen or in winter.

What is Alpine style? - It is where small groups, often just two or three climbers, use an agent to arrange permits and porter support to the base camp. From here, they are self-sufficient and climb unsupported, which means no fixed ropes, no fixed camps, no oxygen and no porter support on the climb.

Where did it start? - The big walls of the Alps and Dolomites were ascended from the base to the top with no support - no fixed ropes. The climbers were self-sufficient, independent and carried all they needed - be it a one day or a multi-day ascent.

From here, the ethic and style spread throughout the world. The inauguration of Alpine style climbing in the Himalayas could have been the spartan 1957 Austrian expedition to the Karakorum, led by Kurt Diemberger. They had no base camp support, no high altitude porters, no oxygen and no fixed ropes. All four members summited Broad Peak, 8047m. Or was it born when Peter Habeler and Reinhold Messner made their incredible three-day ascent of Hidden Peak in 1975? Or.....

Alpine style is The Grail - that mountaineers of substance aim to attain. Historically speaking, once all the world's fourteen 8000m peaks had been climbed, they turned their attention to new, and usually, more difficult routes on the same peaks, or to lower virgin peaks, attempting them alpine style.

Does this mean one has to choose a virgin line or a virgin peak or attempt extreme things to be regarded as a mountaineer or for your trip to be classified as an expedition?

The answer is no. It can be a small group of friends with general mountaineering competence, who will try their first 6000m peak together. They can use a professional western guide and/or a Sherpa. It's all about being honest - with yourself, with others and to the spirit of mountaineering.

Question: *'Then why the controversy - why the criticism of today's big expeditions?'*
Answer: *'It's a question of values, of style and integrity - before, during and after.'*

Mugging old ladies - Back in Norway climbing guru Marius Morstad read my proposed article, *'The decline of Himalayan mountaineering'*.

He came up with an acidic retort, *'You have zero credibility. You have never been above 6500m. How can you criticise those who climb the 8000m peaks? No one listens to a no namer.'*

'True, but I haven't mugged any 80 year old ladies either.
Yet I know it's wrong!'

His words had an irritating ring of reality about them. My interest and concern as to the steady erosion of values in mountaineering was re-awoken by reading *High Crimes* by Michael Kodas.

He sparked the decision that the present downslide to irrelevancy and mediocrity in mountaineering needed to be addressed in some manner. As did the manipulation of the tabloid media at home by Norwegian mountaineer Jon Gangdal - in a brutal attack of Jarle Traa, who had attempted twice to climb Everest, alone and with no oxygen. Traa, believed to be dead, was hung out as being 'irresponsible', where the big 'package-expedition'-style was promoted as the only responsible manner to climb Everest. Traa, like Jesus, came back from the dead (after having spent 3 nights above 8000m, with no tent). Swede Fredrik Stang also blasted the mass media, in a 'self-promote' blitz around the 2008 K2 disaster. Tabloid mountaineering journalism was becoming a world wide phenomenon.

The deeper I delved, the more complex the subject became. There were tales of stealing equipment and food from high camps, of people manipulating photos from other expeditions to show that they had actually climbed a specific mountain. Accounts of Sherpas demanding higher pay high on the mountain, of pre-payment before evacuation etc, all trickled in. There were stories of dry oxygen mixes being sold in advance, only to be swapped to bottles with cheaper humid-air oxygen once the expedition had begun. Lives were put in danger, rescues followed, resulting in more tales of 'non-success' on the killer mountain. Helicopter rescues were daily occurrences, and big business, with guides and agents getting 40% commissions - all paid for by the unsuspecting insurance companies.

There was a story of Sherpa guides leaving a woman client for dead on Kanchenjunga. Other climbers found her alive and took her down. Tales of false claims before, during and after expeditions.

The list grew.

But this was not new. Reinhold Messner, in 1971 had written about similar developments in big wall climbing in the Alps - where nothing was sacred, the goal being: reaching the top, by any method, at any cost.

Chapter 4

The Murder of the Impossible
by Reinhold Messner

What have I got against 'direttissimas'? Nothing at all; in fact, I think that the 'falling drop of water' route is logical so long as the mountain permits it. But sometimes the natural line wanders from the direct line, and then we see climbers, on a first ascent, going straight up as if it weren't so, drilling in bolts. Why do they do that? 'For the sake of freedom', they say.

They have a horror of deviations. 'In the face of difficulties, logic commands one not to avoid them, but to overcome them,' declares Paul Claudel. That is what the 'direttissima' protagonists claim they do, while knowing that today's equipment will get them over any obstacle. Could the mountain stop them with unexpected difficulties? They smile: those times are long gone! The impossible in mountaineering has been eliminated, murdered.

The spirit of direttissima that guides them has infiltrated the entire field of climbing. The climber on overhanging rock drills another hole above the last. He won't give up. Stubbornly, bolt by bolt, he goes on. His way, and no other, must be forced up the face. Expansion bolts are readily used when other methods fail. Today's climber does not want to cut himself off from the possibility of retreat; he carries his courage in his rucksack. Rock faces are no longer overcome by climbing skill, but are humbled by methodical manual labor. Free, naturally protected routes are dangerous so they are over bolted. Goals are no longer achieved by skill, but by overuse of equipment and time.

Times change, as do concepts and values. Faith in equipment has replaced faith in oneself. Today every single limit has vanished, been erased. The word 'impossible' has been eliminated from mountaineering vocabulary - the dragon is dead, poisoned.

Progress? Vast numbers no longer care where they place bolts, whether on new routes or classic ones. Drilling more and more, climbing less and less. Many have lost their taste for real climbing: why dare, why gamble, when you can proceed in perfect safety?

Be cunning, be successful: use every means to 'conquer' the mountain. The mountain can not run away.

Anyone daring to take a stand against current practices is ridiculed. Yet, if we do not oppose, then we are accomplices of the murderers. When future mountaineers open their eyes and see what happened it may be too late: the impossible will be buried, rotted and lost forever.

'They' are ambitious - and manipulate the media - which will inform us that, 'Man has achieved the impossible'. Yet again!

Ask, Is a guaranteed 'impossible' mountain a worthy grail? No, on the contrary, up there we want to find long, hard days and bivi-nights, and mornings when we do not know what the evening will bring.

I am worried about that dead dragon. So I invite you to strive to reach the summits by other routes than those manufactured by man. It is time we repaid our debts and searched for the 'limits of possibility' - for we must have such limits if we are to use the virtue of courage and skill to approach them.

We must never break them down again, especially when it's impossible to reach them.

So let us save the dragon. Let us follow the road that past climbers marked out. Walk in their shadows. Put on your boots and get climbing - take the basic equipment - no more. I'm on my way, ready for anything - even for retreat if I meet the impossible. I will not kill any dragons, join me, we will go to the top together on routes we can do without branding ourselves as murderers.

Mountain Magazine, no. 15, 1971. Translated by Victor Fry.
Revised by DD, 2014.

Book 5

MOUNT EVEREST

An incomplete history of the world's highest mountain

Mt. Everest from 'Peak 22,740ft' - 1935. Photo: L.V. Bryant.

Chomolungma. *Painting by: Tenzin Norbu Gurung.*

A Goddess

She is majestic and she is powerful.
She is honoured and loved by the people
who live in her shadow.
Mountaineers came, saw, and 'conquered' her.
Yet it is they themselves they met.
Their inner fears and tribulations
and
their irrelevancies.
They do not revere her.

Swami Kailash

Chapter 1

A brief glance at point 8848m

The southern flanks of Peak XV, now called Mt. Everest, lie in the Khumbu region of Nepal, while her steep northern and eastern precipices sweep down to the plains of Tibet.

Everest is a western imperialistic and masculine label, named after the Surveyor General of India, George Everest. In the countries of her abode she is referred to in the female gender and honoured with great reverence. In Tibet, she is known as *Chomolungma*, The Mother Goddess of the Earth, Mother Goddess of the Universe, Mother Goddess of the Snows, Mother Goddess of the Winds. Nepal's eminent historian, the late Baburam Acharya, penned the Nepalese name *Sagarmatha*, meaning the Head of the Earth touching Heaven.

She, herself, is a magnificent pyramid of sedimentary rock and metamorphosed limestone covered by veils of snow and ice, and touched by swirling winds and mists. Her foundations were laid deep below a primal ocean over 450 million years ago. Mother Earth was restless, which resulted in the division of Gondwana, a lost continent that lay in the far south. As Gondwana broke up the different sections formed what were to become today's continents of Antarctica, South America, Africa and India.

The newly created India drifted northwards at the speed of 15-20cm per year. On reaching the equator, after about 50 million years, she slowed down to about 5cm per year, which is twice the speed that human fingernails grow. Early in the Jura period, the land masses collided and the ocean floor raised itself to form the Himalaya chain. This stretches from Kashmir to Assam. Rivers flowed, plants grew and animals roamed. Aeons passed, and glacial, wind and water erosion continued to sculpt. In addition, there were the ever-present mighty Gods and Goddesses, good and evil spirits, all contributing to the formation of the Himalaya.

In a second of micro-time man entered the scene, and walked below the shadow of the mountains, both in fear and revering them. Man settled along the banks of seven major rivers: the Yangtze, Yellow, Indus, Mekong, Ganges, Salwen and Yarlung. All of which start their journey from Himalayan snow and glacier melt. On Everest's northern side lies the great Tibetan plateau, often called The Roof of the World. From here, over 2000 years ago, empires grew and dwindled, with migration from north and south, east and west. A rich kaleidoscopic of cultures.

On the Nepal side, Neolithic tools have been found that originate 9000 years back in time. Today, approximately 47 different peoples live in Nepal, some say over 200 - of which the majority are Hindus.

Nepal is a rich geographical tapestry. In the south are the Terai-Gangetic plains, the lowlands that border India and the most highly populated area. Then we move north to the river rich Pahad region, the so-called foothills that rise from about 800m to 4000m. In the northwest of the country, there are hidden and rarely visited alpine valleys and a high mountain desert area, Dolpo, which merges into Tibet. Along the backbone of Asia, are the fourteen Himalayan giants that rise to over 8000m - eight of which reside in Nepal.

Up until the beginning of the nineteenth century, it was generally believed that the Andes were the highest mountains in the world. In 1804, Captain Charles Crawford, was posted to Kathmandu, and produced a map showing, *'the whole of Nepal, with a line of 'snow mountains'*. In 1808, The British India Survey began producing reliable maps to the entire Indian sub-continent - a daunting task. It took 30 years before they reached the Himalayas, but pre-empting them, in 1810, Lieutenant W.S. Webb set out to trace the source of the Ganges. In the process he calculated 'some peaks' to be in excess of 7925m. If true, this would make them the highest in the world. However, geographers outside India dismissed his observations as 'utter nonsense' - and the Andes continued to rule supreme.

Interest in the Himalaya was great and not just for mountaineering purposes. In 1847, The Great Trigonometrical Survey of India undertook a set of triangulations of Nepal with measurements being taken from India. Most of the peaks were allocated a number or a letter to identify them. At that time, Kanchenjunga (8598m) was regarded as the world's highest mountain. Peak XV (Everest) was virtually ignored, as it was thought to be lower than its neighbours, Lhotse and Nuptse. It was during the 1847-55 study, that Peak XV received its official status as the world's highest mountain - at 8840m.

Colonel Waugh, who led the study, recommended to the Royal Geographical Society that the peak be named in honour of Colonel George Everest, his mentor (later knighted, Sir G. Everest). Everest himself preferred the use of numbers or local names, noting that local people had difficulty in pronouncing western names. He had never seen the mountain, declined the honour, and died the year it received her official name: his.

Colonial British paranoia feared that the Czar of Russia was setting his eyes on the Jewel of the Empire, India, with intent to invade via Tibet. Britain moved first and a new shameful page of British colonial history was written: the invasion of Tibet*. In 1905, when Francis Younghusband withdrew his conquering army, Captain C. G. Rawling viewed the North Ridge of Everest from 100km away and felt, *'It might provide a feasible route to the summit.'* He was proved correct when, 55 years later, Chinese mountaineers made a difficult yet successful ascent by this line.

The Golden Age of European mountaineering - was fading, as all the main peaks in the Alps had been climbed. Attention was being directed towards the Indian Himalaya, because neither Tibet nor Nepal allowed expeditions to cross their boundaries. J.B.L. Noel (John Baptist Lucius), dressed as a Tibetan pilgrim, entered Tibet illegally and came within 50 miles of Everest in 1913. His reports were greeted with great interest in mountaineering, geographical, colonial and in military circles. In 1921, the Tibetan authorities granted permission for the first British expedition to enter the country. This was led by Charles Kenneth Howard Bury, and included George Herbert Leigh Mallory, whose name will always be associated with the mountain. After a long march from India, Mallory wrote:

'As the clouds rolled asunder before the heights, gradually, very gradually, we saw the great mountainsides and glaciers and ridges, now one fragment, now another, through the floating rifts, until, far higher in the sky than imagination dared to suggest, a prodigious white fang - the summit of Everest appeared.'

They discovered the link between the East Rongbuk Glacier and the North Col - the route that would be used by all expeditions for the next two decades. They also established a camp at 6800m and gained the Lho La - a pass that allowed them to look into Nepal and discover what would later be called The Western Cwm.

Cwm, being Welsh for basin, was probably adopted by one of the early expeditions who had trained in the mountains of Wales.

Mallory, Bullock and Wheeler reached the North Col, at about 7010m. From here, Mallory contemplated a push to the summit. Fearful blizzards and exhaustion put a stop to this, but Mallory was certain they had unlocked the secret to the summit.

*Suggested reading, *Bayonets to Lhasa* by Peter Fleming

The British - returned the next year, led by General Charles Bruce, full of optimism. However, Mallory sounded a cautionary note, *'We must remember that the highest of mountains is capable of severity, a severity so awful and so fatal that the wiser sort of men will do well to think and tremble even on the threshold of their high endeavour.'*

His words rang true, because by May 20 half the porters were suffering from altitude sickness and the mountain was swept by biting winds. They set off from Camp IV at 6980m, and established Camp V at 7620m. Mallory's fingertips and Norton's ears had succumbed to frostbite, and Morshead's toes and fingers were in a bad way. Next morning, Mallory and Norton continued, and reached a height of approximately 8350m, using supplementary oxygen.

A second and then a third attempt were made. During the latter, four climbers and 13 Sherpa set off from Camp III, reaching a point just below the North Col - where they were struck by an avalanche.

Mallory, *'...a noise not unlike an explosion of gunpowder. ...the snow's surface broken ...I instinctively moved. A wave of snow came over me. ...the matter was settled. ...then I was moving downward ....somehow I managed to turn out from the slope so as to avoid being pushed headlong and backwards ...then the rope at my waist tightened and held me back. However, I thrust out my arms to keep them above the snow and at the same time tried to raise my back. Beneath the surface of the snow, with nothing to inform the senses of the world outside it, I had no impression of speed. I struggled in the increasing pressure about my body, wondering how tightly I should be squeezed, and then the avalanche came to rest.'*

Mallory found himself on the surface, dazed. The avalanche had swept away seven Sherpas, and Everest had claimed its first victims.

Mallory and Irving - Two years later, in 1924, the British returned for the third time.

General Charles Bruce was the leader, only to be struck down by malaria on the gruelling trek across Tibet. He was replaced by Major Edward Norton, and Mallory was appointed climbing leader.

Mallory had settled into a new job and was wary of being seperated from his family, yet again. He still harboured guilt over the loss of the seven Sherpas, which he recalled as being, *'brave men,'* whom were, *'ignorant of mountain dangers, like children in our care.'* He had not been openly criticised for choosing to traverse the North Col, so soon after fresh snow, but there were mutterings. He himself accepted his share of blame, noting, *'More experience, more knowledge might perhaps have warned us not to go there.'*

Again high winds and heavy snow hampered their efforts. On descending to Camp III, Mallory fell into a crevasse, *'The snow gave way ...snow tumbling all round me, luckily only about ten feet ....half blind and breathless to find myself most precariously supported only by my ice axe, somehow caught across the crevasse - and below a very unpleasant black hole.'* Worse was to come, with a harrowing two day rescue of porters, frostbite, heavy snows and storms.

1924 Mount Everest Expedition. *Base Camp. Back row (L to R): Andrew Irvine, George Mallory, Edward Norton, Noel Odell and John Macdonald.*
Front row: (L to R) Edward Shebbeare, Geoffrey Bruce, Howard Somervell and Bently Beetham. Photo: J.B. Noel © Royal Geographical Society (RGS).

With true British fortitude, the expedition continued. Somervell and Norton made another attempt, until Somervell was totally exhausted and had to admit defeat. Norten continued alone, reaching what was to be named the Norton Couloir. *'I began to feel that I was too dependent on the mere friction of a boot nail on the slabs. It was not exactly difficult going, but it was a dangerous place for a single unroped climber, as one slip would have sent me to the bottom of the mountain.'* He achieved a new height record of 8575m - alone and without oxygen.

Mallory and Andrew Irvine were to make a further attempt. Supported by Odell, the expedition's geologist, they passed Norton's high point. From below Odell noted, *'The entire summit ridge and final peak of Everest were unveiled. My eyes became fixed on one tiny black spot silhouetted on a small snow crest beneath a rock step (now named The Mallory Step) in the ridge, the black spot moved. Another black spot became apparent and moved up the snow to join the other on the crest. The first then approached the great rock step and shortly emerged at the top, the second did likewise. Then the whole fascinating vision vanished, enveloped in a cloud once more.'*

Odell was the last person to see Mallory and Irvine alive, as they entered the summit cloud, that, *'remote and inhospitable spot'* and never returned. The body of Mallory was discovered by Wang Hung-no in 1975 and left undisturbed. In 1999, it was re-discovered by Conrad Anker but there was no sign of his camera. The question if they had reached the summit remains an open one. Let it remain so.

George Finch testing oxygen system. Photo: J.B. Noel. © RGS

Sherpa carrying load. Photo: J.B. Noel. © RGS

Something calls - 1933 saw a new British team led by Hugh Ruttledge. The highest camp ever established was pitched on a small sloping ledge at 8350m. From here, Laurence Wager and Percy Wyn-Harris continued the next day to reach 8570m. This was the approximate height Norton had gained in 1924. Here, they turned back due to poor snow conditions and the lateness of the hour.

The Mad Yorkshireman - Next year, Maurice Wilson, who harboured dreams of being the first man into space, flew a small plane to India, where he was denied entry to Tibet. Rather than return home, he sold the plane, disguised himself as a monk, and walked 400km to reach the northern side of the mountain. With no ropes and no proper equipment his porters went on strike. He insisted that they should continue. They tried to dissuade him, using both arguments and force, but he refused. Everest-madness had set in. He took a tent, three loaves, two tins of oatmeal and a camera - plus a silk Union Jack. Telling the Sherpas to wait for two weeks, and pointing towards the summit, he set off alone. The last entry in his journal was, *'Off again, gorgeous day.'* His body was later found on an approach to the North Col and, according to his journal, he had reached a height of 7560m (disputed).

There were further British attempts. One in 1935, led by Eric Shipton, where they tried climbing during the monsoon period - which proved too dangerous due to storms and to the frequency of avalanches. Shipton was convinced it was possible to reach the summit via the West Col - but this would require entry via Nepal, which was closed to foreigners.

The 1936 expedition, led by Ruttledge, was thwarted by continuous new snow and avalanche danger. Shipton and Wyn Harris were swept away by an avalanche that luckily stopped 100 metres from the edge of a steep cliff. Again, an early monsoon stopped play. The 1938 expedition, led by Bill Tilman, adopted Shipton methods: 'small and light'. However, they were plagued by illness, and the storms of an early monsoon stopped them at 8300m.

World War II put the mountain on hold. The next attempt had to wait until 1947, when Canadian Brit Earl Denman, together with two Sherpas, one of whom was Tenzing, made an illegal attempt. It was poorly planned and poorly equipped and as such, doomed to failure. If successful, the ascent would not have been officially acknowledged. Therefore I asked Tenzing why he had taken the chance, he smiled,

'It was Everest.'

Tibet closed its borders, a decision based on a horoscope reading that warned the Dalai Lama to be wary of tourists who were seeking treasures in the 'home of the gods'.

Luckily for mountaineers, in 1949/50, Nepal removed its restrictions. The British quickly fielded two reconnaissance expeditions, one in 1950, an Anglo-American affair, which reached the foot of the Khumbu Icefall. The next year, led by Shipton, a route was established up the vast amphitheatre between Everest and Lhotse to the lip of the Western Cwm. Here to be stopped by 30m wide crevasses. As compensation, they had discovered a route that in all probability, would lead to the summit. They also brought back viable proof of the existence of the Yeti - photographs of Yeti footprints in the snow.

Up to this point the British had virtually monopolised the mountain.

A Danish traveller, Klavis Beker Larsen, who had never held an ice axe, recruited a number of Sherpas, and reached the North Col, only to be turned back by rock fall.

In 1952 the Swiss - made a strong attempt, with Tenzing as leader for a Sherpa team of twelve. Tenzing had added Norgay to his name, meaning, 'the fortunate one,' or 'he who follows the gods'. They were the first people to step foot into the elusive Western Cwm, from where they climbed the Lhotse Glacier and traversed to the South Col - preparing for a final push. A combination of illness and frostbite had taken its toll, with the other Sherpas being unable to continue. Hence, three Swiss climbers and Tenzing were left to carry the heavy loads.

At just under 8400m the four set up a tent bivouac, with enough room for two. Lambert and Tenzing would spend the night, with no primus or sleeping bags, and make a bid for the summit the next day. *'Our legs would not obey us and our brains scarcely functioned. Our hands worked more skilfully without gloves, but to take them off would cost us dear.'* The wind was relentless, and the cold seeped into their bones. To sleep would have invited the inevitable frostbite, so they spent the night shaking each other awake. Lambert recounts, *'I dare not sleep, must not sleep. The stars were so brilliant that they filled me with fear.'* With no food or stove they resorted to melting a lump of ice over a solitary candle. They started at 06.00 the next morning, only to be greeted by appalling conditions that slowed them down. They took nearly six hours to climb less than 200m, over reasonable easy terrain. With less than 250m, in height gain, to the summit their leaden bodies screamed, time to turn around.

A difficult descent followed, to re-join their exhausted companions, and call off the expedition. Lambert is one of the sports least known 'greats'; and the holistic manner in which he conducted his lifelong approach to mountaineering is worth emulating.

However, the Swiss had a second expedition planned that year, returning after the monsoon. Things went well until a giant ice block thundered down the Lhotse face, killing one Sherpa, injuring and sending three others 200m helter skelter down the face.

The Third Pole - Both the North and the South Poles had been reached and the French had climbed Annapurna, the world's first 8000m peak to succumb to man. With the Swiss nearly reaching the top of what was now referred to as The Third Pole, the race had become political and nationalistic.

The next British expedition was in 1953, and was led by Col. Henry Cecil John Hunt, who had unexpectedly replaced the vastly experienced Eric Shipton as leader. Shipton's idea of an expedition was, *'two men, a dog and a plan on the back of an envelope.'* This 'gung-ho' approach did not suit the political powers in Whitehall.

The British Empire needed The Third Pole.

John Hunt was a military man with wide experience in running large, complex organisations, and his credentials were superb. In addition to having climbed in the Alps for ten seasons, he had been on five Himalayan expeditions.

Griffith Pugh - unsung hero of Everest - Historically, it has been portrayed that the previous British expeditions had been led by and comprised off the cream from British mountaineering circles. Harriet Tuckey's book, *Everest The First Ascent - the Untold Story of Griffith Pugh, the man who made it possible,* opens an inconvenient pandora's box. Chris Bonington notes, *'Shines an entirely new light on the great expedition, a riveting read, full of surprises.'*

There is no doubt that Hunt provided meticulous planning, great depth of knowledge and he led from the front, reaching a height of 8335m himself. On the negative side, he and the upper ranks of the establishment, hampered the work of Dr. Griffith Pugh, ridiculed and side lined him and attempted to remove his contribution from the history books. Pugh, an eccentric scientist (physiologist), did not suffer fools easily and lacking diplomacy he had ruffled a few feathers. He clearly pinpointed the issues that had hampered the early expeditions - and devoted his energy in trying to rectify them.

They had included poor leadership, vague communication, haphazard planning, poor logistics, poor individual and group hygiene, leading to sickness, poor diet, poor and often insufficient equipment, plus a limited knowledge of the pathology of acclimatisation. All these, and more, had contributed to failure after failure - where avalanches and weather had been mooted as the reasons.

Griffith Pugh, scientifically, systemised the whole process of expedition mountaineering. He observed, registered and deducted, plus he learnt from others - not least from the French and Swiss, who were in the main professional guides. He formulated menus suitable for life at altitude, put personal and group hygiene high on the agenda, and developed a whole range of equipment. This ranged from high altitude boots, clothing, sleeping bags and tents, to various combinations of materials with different functional performance values. Not least, he documented altitude height gains in relation to fitness, acclimatisation adaption, deterioration and regeneration - both physical and psychological. He ran parallel tests on both the open and closed oxygen delivery systems, leading to increased understanding and improvements. He, against a powerful establishment stuck in the past, contributed greatly to the successful first ascent of Everest and to the development of Himalayan mountaineering.

12 February 1953 - Colonel Hunt and his party set sail from Britain aboard the S.S. Stratheden, bound for India. From there they flew to Nepal, where they recruited 350 locals to carry the expedition equipment to the monastery at Thyangbochie, at 3950m. Here they spent two weeks experimenting and acclimatising, before freighting the equipment to base camp, on the Khumbu Glacier.

Georg Craig, in his 60th Anniversary book, *Everest*, notes, *'Hunt was concerned about the Khumbu Icefall, which was a giant glacier sporting a multitude of crevasses, ice walls and tottering ice pinnacles. They hacked steps out of the ice to facilitate the passage of the laden Sherpas, wriggled through narrow cracks, fixed ropes and suspended ladders as bridges to reach the half way terrace. Sometimes they would climb up in the morning to find that the previous days tracks had been obliterated by an overnight fall of ice.'*

On reaching the upper part of the icefall Hunt wrote, *'...the steep slope on the far side was covered by blocks of ice of all sizes, piled in indescribable confusion on a wide front and extending over 60m up the slope. The collapse of any one would have spelt disaster to a party below ...exciting climbing and always underneath, we had a feeling of tension and danger.'*

From here, it was a question of establishing camps, ensuring they were filled with supplies, to ensure that all was 'in place' for the all important summit bid. Bourdillon and Evans were to make the first attempt - with the South Summit as their main goal. If the way to the main summit seemed reasonable, and time allowed, they were 'authorised' to make an attempt.

Time did not allow them this honour, but they did invaluable work in preparing the way, as well as leaving a cache of oxygen bottles, for the second team, Hillary and Tenzing.

*Edmund Hillary and Tenzing Norgay on South East Ridge, 1953.
Day before summit attempt. Photo: Alfred Gregory © RGS*

Hunts organising ability and the teams last day's efforts had positioned Hillary and Tenzing at 8500m, just 300+ vertical metres below the summit.

That night they dined on dates, sardines, biscuits, jam, honey and tinned apricots, thawed out over the Primus. They had enough oxygen for four hours use each that night, and could doze 'comfortably'.

At 06:30, they crawled out of the tent, connected up the oxygen apparatus and set off in the early morning light. Still concerned about his cold feet, Hillary asked Tenzing to lead the way. Far above them, they could see their first objective, the South Summit. As Hillary's feet warmed he took over in front and they moved steadily along the ridge. The snow was only marginally predictable, suddenly giving way without warning, all of which was very unnerving. They reached the south summit 09:00.

Before them lay the virgin ridge that led to the summit itself, a daunting sight, with giant drops on each side. Especially alarming were the huge overhanging masses of snow and ice, which stuck out like twisted fingers over the 3000m drop of the east face. If they accidently set foot onto these cornices, it would spell disaster. Hillary calculated that they had four and a half hours of oxygen each so, with deep breaths, they stepped into the void.

To their relief the snow was firm, and they were able to cut steps and gain secure ice axe anchors. Straddling the ridge, they could see the tiny tents of Camp IV 2500m below, nestled in the Western Cwm. An exhilarating position, with both climbers feeling on top form, clad in excellent equipment - and the highest point in the world before them. After one hour, they reached a steep section, the last obstacle, and seemingly unpassable.

On the east side, Hillary found a line of weakness in the final defence, and, *'I jammed my way into this crack, then kicking backwards with my crampons, I sank their spikes deep into the frozen snow behind and levered myself off the ground. Taking advantage of every little rock hold and all the force of knee, shoulder and arms I could muster, I literally cramponed backwards up the crack, with a fervent prayer that the cornice would remain attached to the rock.'*

It did. Tenzing followed and they plodded on to reach the summit at 11:30 together. Tenzing recalls, *'...my mountain did not seem to me to be a lifeless thing of rock and ice, but warm, friendly and living.'*

Tenzing Norgay. Photo: Ed Hillary © RGS

The British Mount Everest Expedition was a success, with Hillary and Tenzing reaching the summit on 29 May, 1953.

The British Empire's honour had been saved - The Third Pole was theirs, and the news was relayed back in time for Queen Elizabeth's Coronation. Yet neither of the two summiteers was British, Tenzing was of Tibetan descent and Hillary, a beekeeper from New Zealand. This technicality was drowned in the media blitz that followed and the conquest became an integral part of the Queen's Coronation.

'Because it is still there.' Victor Kienov.

From the first attempt in 1921, to the first successful ascent in 1953, fifteen people had lost their lives. Now with Everest conquered many assumed interest would diminish - the cost in time, energy, prestige, money and risk to life and limb was too high. But no, interest increased.

Hillary foresaw this, *'I remember looking over the Himalaya and not feeling that I had closed the door on exploration, but rather just the opposite, I remember thinking, 'God, the possibilities are endless!''*

It was fitting that the second ascent was made by a Swiss party, via the same route, in 1956.

But Hillary had noted, *'...the possibilities are endless.'* Meaning the potential for new and more demanding routes on Everest, and on other mountains in the Himalaya.

The Chinese took up the baton and, in 1960, an expedition led by Zhang-Chun Shi, comprising of 214 members, followed the line of the earlier British attempts up the North Col. They successfully placed three members on the summit, using five weeks. Their success was credited to the inspirational words of Chairman Mao! Considering that many of the climbers were relatively inexperienced, that one took off his boots to negotiate a difficult passage, that they had run out of oxygen, and that they reached the summit after dark, raised doubts. As more information emerged, e.g. the barefoot climber showing his toeless feet lost to frostbite and collaborating photographs, saw this difficult route being rightfully credited as the first complete ascent from the Tibetan (north) side of the mountain.

Breakthrough - In 1963, the West Ridge and the first traverse of Everest were the goals of Americans Tom Hornbein and Willi Unsoeld.

Joseph Poindexter writes, *'They set out from their tent high on the West Ridge over ground that had never been touched by man. Thirteen hours later, they arrived at the summit. A new route, but the most dangerous part was just beginning. The pair planned to complete the first ever traverse of Everest by descending to the South Col. They hoped to join Lute Jerstad and Barry Bishop, who had summited earlier the same day by the normal route. But at 18.35, it was already desperately late. They followed the other's footprints until the light failed.'*

The alpine-style climbing to the summit had been technically difficult, route finding complex, and they had little chance of descending should things have gone wrong. Something Horbein and Unsoeld must have been very aware of.

From Hornbein, in *Everest: The West Ridge*, *'Too much labour, too many sleepless nights, and too many dreams had been invested to bring us this far. We couldn't come back for another try next weekend. To go down now, even if we could have, would be descending to a future marked by one huge question: what might have been? It would not be a matter of living with our fellow man, but simply living with ourselves, with the knowledge that we had had more to give.'*

Now, with the mountain cloaked in darkness, they had two alternatives, dig a hole and spend the night with the potential danger of odemina or frostbite, or continue to descend.

They chose the latter and caught up with Jerstad and Bishop. Eventually the four had to stop, exhausted, *'...with no more protection from the inhuman cold than their down jackets, huddled on the rocky ridge to await the dawn.'* Unsoeld suffered from frostbite, lost toes and fingers, but their ascent was a major achievement - ranking with the finest in Himalayan history.

Mt. Everest saw the first attempt to ski down in the spring of 1970, when Yuichiro Miura skied from the South Col (7925m), with a parachute as an 'air brake'. As he skied, all he could hear was silence. *'I am alone in a world without sound. Because the air is so thin and the wind is at my back, I feel nothing, like a rocket streaking through vacuous space.'* Within six seconds, his speed had reached nearly 200 km an hour. To counter this, he opened his parachute, but, *'Is it the wind or the thinness of the air? There is nothing for the chute to hold. It drags along uselessly behind me.'* He hurled towards oblivion, twisted and turned on his skis trying to slow down, lost control, resigned himself to death, then he hit a rock and made one of history's most spectacular falls. Remarkably he survived.

Women on Everest - In 1924, Anne Bernard had requested a place on the next British expedition to Everest, to which the Everest Committee replied, *'Impossible to contemplate the application of a lady to take part in a future expedition to Everest. The difficulties would be too great.'*

Attitudes towards women in Japan are not so different, where custom forbids a woman to leave her husband without his prior consent. Junko Tabei was invited to Everest, but her husband insisted she give him a child first. She did her duty, and then went on to prove the Everest Committee wrong, in 1975, by being the first woman to reach the top. Eleven days later, Phantog, a Tibetan member of the big Chinese expedition, noted earlier, was the second female to stand on the top (losing three toes in the process).

Southwest Face - The highlight of 1975 was Bonington's expedition that climbed the impressive and difficult Southwest Face. Six expeditions had already tried and failed to find the key to this giant face. Bonington's party reached what was called, the Great Central Gulley at 7925m, to be stopped by a menacing black cliff. This in turn was passed by Nick Escourt and Tut Braithwaite - who, late in the day descended, making way for a second team, comprising of Dougal Haston and Doug Scott. They followed a succession of ramps and gullies to the South Summit as the sun was setting and where they ran out of oxygen.

Even so, they continued, and summited at 18:00hrs.

Dougal Haston on the Hillary Step, 1975. The first ascent of the South West Face. Photo: Doug Scott

They started their descent at 19:00, but did not get far in the darkness. With no bivouac equipment, they scraped out a hollow in the snow to spend the night. They lost neither toes nor fingers, nor seemingly any brain cells, but they did hallucinate.

Doug remembers, *'Dougal had a long conversation with Dave Clarke, who was 2000m below us in the Western Cwm. From the way Dougal was talking, it was as if Dave was here in the cave with us. They were discussing the relative merits of various sleeping bags, obviously a manifestation of Dougal not having one. I found myself having conversations with my feet as if they were two separate, quite distinct personalities right there in the cave.'*

The expedition was important, in that they were raising standards, but unfortunately, it was marred by a tragedy. The second summit team, Peter Boardman and Sherpa Pertemba were descending from the top when they were surprised to meet cameraman Mick Burke, on his way up. His plan was to film them on the summit. However, with Pertemba losing feeling in his fingers and toes, they had no option but to continue the descent. Mick, now so close, continued to the summit, alone. He was never seen again.

Everest without bottled oxygen - The first ascent without artificial oxygen was in 1978, by Peter Habeler and Reinhold Messner - two of the world's leading climbers. They both had climbed some of the most difficult routes in the Alps, in good style and at speed. Plus they had climbed Gasherbrum, the 11th highest, without oxygen.

Everest was next. Their ascent was hampered by deep snow that both slowed and exhausted them. As the storm intensified they reached the south summit, and stood above the clouds, too tired to talk, they resorted to gestures.

'I felt somehow light and relaxed, and believed that nothing could happen to me. At this altitude, the boundaries between life and death are fluid. I wandered along this narrow ridge, and perhaps for a few seconds I had indeed gone beyond the frontier which divides life from death - physically finished. I was no longer walking of my own free will, but mechanically, like a robot. I seemed to step outside myself, and had the illusion another person was walking in my place.'

Habeler had experienced an out of body situation, jolted back to life by an extreme cramp in his fingers. They pushed on, and ended up practically crawling, one leg in Tibet, the other in Nepal, to the summit.

Medical prognosis was that anyone climbing at this attitude without bottled oxygen would suffer permanent brain damage. With this in mind, Peter descended the 800m to the South Col in about an hour (a distance they had taken 8 hours to climb up). He glissaded most of the way on the 'seat of his pants'. Messner lingered, savouring being alone at the highest spot on the planet.

The West Ridge Direct was climbed in 1979, by a strong Yugoslav team and was probably the highest grade V in the world at the time. The first winter ascent was by Polish climbers Krzysztof Wielicki and Leszek Cichy, who battled through the Jet Stream winds to reach the summit on 17 February 1980. Three months later, a Japanese expedition completed the first full ascent of the North Face. Another Polish team, in the spring of the same year climbed the South Pillar of the SW Face, another new route (by Kukuczka and Czok).

Alpine style solo - In 1980, during a lull in the monsoon (mid. August), Reinhold Messner made the first alpine style solo ascent, without bottled oxygen, up the North Col and Great Couloir. He carried a 44kg sack, with clothing, stove, fuel and food for seven days. Totally alone, he crossed vast areas of crevasses without rope, each step a tentative one with potentially catastrophic consequences should a snow bridge collapse.

One did. He plunged into a crevasse, *'The sweat froze in my hair and beard, but the anxiety in my bones disappeared the moment I started moving. At each movement, however, the feeling of falling again came over me, a feeling of plunging into an abyss, as if the ground was slowly giving way.'*

He climbed out of the crevasse, composed himself and continued up the north ridge, to set up a camp at 7700m. From here he climbed via the north east ridge, but was then forced to traverse into the north face to reach the Norton Couloir - only gaining 400m that day. A new bivouac, not as high as hoped and time was not on his side. The next morning he left his tent and most of his equipment to make a lightweight push for the summit. His gamble worked - by moving fast and free he reached the top at 15:00. - *'I squatted down, feeling as heavy as a stone. I just wanted to rest and forget everything. I was leached, completely empty.'*

A milestone had been passed in world mountaineering.

The first Russian ascent went, via a new and difficult route, up the Central Pillar of the Southwest Face in 1982, putting eleven people on the summit. Then in 1982, Pete Boardman and Joe Tasker attempted one of the last great challenges on Everest, the enormously long Northeast Ridge, alpine style. The ridge is adorned by three giant pinnacles, between 7900m and 8380m, offering one of the last great classic lines on Everest. Tragically, both disappeared around 8200m.

The route was later climbed by Harry Taylor and Russell Brice in 1988, in terrible conditions. On having completed the ridge they did not go for the summit but descended, when joining the normal route. Three years later, Boardman's body was found by climbers from Kazakhstan, in a sitting position, *'looking like he was asleep.'* Then in 1996, Japanese climbers came across the body of Tasker, by the second pinnacle.

Their having spent over three days in the so called death zone, without oxygen, coupled with difficult climbing, had taken its sad toll.

In 1983, a large all star American team attempted the east, or Kangshung Face. They used 28 days to climb the main buttress, and once above this major obstacle, they still had over 2200m of exposed climbing, over avalanche ridden slopes to contend with. Some of the expedition members had descended in view of the objective dangers. The remaining six reached the top, only to witness three Japanese falling to their deaths on the south east ridge.

In 1984, a small and not too experienced, Australian team established a new and difficult route on the North Face, without oxygen. In 1986, Erhard Loretan and Jean Troillet made an alpine ascent via the North Face in record time and glissaded back down in just 3½ hours.

Beneath an ice serac on the Kangshung Face.
Photo: Stephen Venables © RGS

In 1988, an Anglo-US group of four pioneered another route on the steep Kangshung Face, with no Sherpa support and no oxygen. A strong team effort, although only one actually summited, Stephen Venables. He then continued to descend to the South Col alone, returning to Kathmandu to phone his parents to let them know he was alright, *'Oh, got to the top, sorry, running out of money for the phone, talk later.'*

Later that year, French ace Jean-Marc Boivin took a paraglide from the summit to the Western Cwm in 11 minutes.

1990 heralded a number of interesting ascents. Australian Timothy Macartney-Snape started from sea level in Bombay and walked 1500 miles before climbing the mountain. In 1995, Alison Hargreaves was the first woman to reach the summit alone and without oxygen. In 1997, Goran Kropp cycled from Sweden, climbed the mountain on his own without oxygen and then cycled home to Sweden. In 2000 Davo Karnicar, from Slovenia, opened the new millennium with the first complete ski descent, following the original Hillary/Tenzing line.

Everest has been the arena of many records. Nepali Min Bahadur was, in 2008, the oldest person to reach the summit, at 76. Jordan Romero, at 13, became the youngest in 2010. In the same year, Appa Sherpa made his twentieth ascent.

In 2012 Ralf Dujmovits - photographed what was to be called 'the human snake' - where about 600 climbers from 39 different expeditions were winding their way up the mountain along a fixed line of ropes - following the line first climbed by Hillary and Tenzing.

'My Father would be shocked.'
- Jamling Tenzing Norgay

The Lhotse Face, on Everest. Photo: Ralf Dujmovits.

A pristine campsite between the pinnacles, Rongbuk Glacier, 1932. Photo: J.B. Noel ©RGS

'She sheds tears, more and more each year.'
Lhakpa Bhoti

Litter at Advance BC on the Rongbuk Glacier. Photo: Roger Mear © RGS

Book 6

Why?

Chapter 1.

Does Why lead to How?

What values lie in the depths of our sport?
How did they develop?
And why do we climb?

> *'It is not **what** you climb - it is **how** you climb.'*
> Swami Kailash.

Norwegian, eco-philosopher, Nils Faarlund disagrees and asks us take this thought a little deeper, **'The decisive point is the answer to the question of Why.'**

Faarlund's standpoint, I believe, is that an honest answer will reflect the ethic the individual, party or expedition has. This in turn lays the foundation, or understanding, as to how the individual or party should or will practice the sport in the mountains - and reflect in their actions afterwards.

Will an answer lead to actions being taken in respectful sincerity to the challenge that nature places before us? Will we grasp and solve the mysteries that nature presents? Or will we act like half mad elephants running amok in a crockery store?

A Lion of Pure Gold - The monks at Rongbuk Monastery on the Tibetan side of Everest believed that the British were seeking, *'The lion of pure gold that resided on the summit.'* What other explanation could there be for their incessant and unexplainable obsession, the use of enormous resources, the danger, the suffering, the uncertainty and the continual loss of lives?

All of which must have seemed absurd to the monks.

Rongbuk Monastery - Lies at 5030m on the north of Everest and was founded in 1902. At the height of its influence, there were over 500 monks, where today there are but a handful. Nearly all seven British expeditions, between the wars, were blessed by the head Lama, Ngawang Tenzing Norbu. General Bruce, in 1922, noted that his holiness was, *'full of dignity, with a most intelligent and wise face.'*

Rongbuk Monastery with Everest in the background. Photo: J.B. Noel © RGS

The area was a sanctuary, and many British expeditions observed with both surprise and pleasure the abundance of tame wild animals and birds. The reason being, that no animal was allowed to be killed in the area. The lamas even offered yaks to the gods, but instead of being sacrificed, they would release them into the wild. The monastery was destroyed by the Chinese during the Cultural Revolution in 1960 and has since been rebuilt. Now, both monks and nuns reside here, as 'tourist attractions' under the watchful eye of the Chinese.

Why do we climb? - The majority of climbers have been asked why they climb. Those who question, often follow up with, *'It is not for me,'* or, *'I do not understand, why you do it,'* or, *'You would not find me up there.'* And they shake their heads.

Just the never ending question and reaction places our sport and our activities in a special light, giving mountaineering a slightly mystifying position - with the consequence that it gains a status that can, and is, easily manipulated by 'the spin doctors' of our sport, for their own purposes. Everest has become a victim.

There has been many an attempt to find a plausible explanation. Conrad Gesner (1516-65), was one of the first to climb mountains for their own sake, stated: *'...marvellous and unaccustomed...dwelling among the clouds...'*

Marie Paradis (1778-1839) and Henriette d'Angeville (1794-1871) were the first women to ascend Mont Blanc. Their ascents were 30 years apart and 'why they climbed' was for two totally different reasons. For Paradis, it was a publicity stunt to enhance her teahouse business, while for d'Angeville, it was a part of the emancipation movement, which evolved into, *'a desire to climb, so ardent that it gave movement to my feet'.* Paradis never climbed again, while d'Angeville completed 29 climbs, the last when she was 69.

'Because it is there.' - The most famous answer came from Mallory in 1923, *'Because it is there!'* On the surface, it may appear as a rather flippant gut-response. Yet, he was a man of letters, a seeker, a thinker, and his reply went deeper, *'It is the struggle of life itself, forever upward. What we get from this adventure is sheer joy.'*

The simplicity of, *'sheer joy',* can that be bettered?

Edward Whymper, whose first ascent of the Matterhorn was the forerunner to modern mountaineering, presents a similar view, *'Out of toil comes strength, an awakening of all the faculties, and from the strength arises pleasure.'* Faarlund's own explanation is, *'Nature is the source and repository of Culture, the essential spark - mountaineering is one road to the source, a way home.'*

Non climber, Mahatma Ghandi tells us of an Urdu saying, *'Adam is not God, but a spark of the Divine.'* Adam symbolising mankind; you and me. Could this 'essential spark' be deeply imbedded, as a part of our essential selves? A soul which is an integral part of the Divine, of God? The eternal spirit that is the soul of the Universe, where all is one and one is all? Can climbing bring us closer to our inner spirit?

If God is nature and nature is God, could climbing undertaken in simple 'purity', a meeting with nature face on, be a meeting with our spiritual selves? Millions have sought and found God in the mountains and the wilderness, including nearly all primal religions.

For Gesner, it too was almost 'godly', for Paradis pure 'business', for d'Angeville, it may have been 'physical euphoria'. Mallory chooses 'sheer joy' and Whymper concurs with 'pleasure'.

Faarlund postulates that mountaineering is, *'serendipitous steep-land art.'* A satisfying thought in that all true art stirs body and mind and has the potential to stir our soul.

Jeff Jackson (in *Ascent*) implies that 'pleasure' and 'sheer joy' cannot be sufficient. He notes that Conrad Anker, Jimmy Chin and Renan Ozturk returned from Mount Meru in 2008 being, *'...utterly depleted after their attempt. Chin went home in a wheelchair and Ozturk couldn't walk for two weeks. Yet they returned in 2011.'*

Where is the pleasure in that?

Jackson summarises, *'Obviously, the reward must transcend mundane definitions of happiness. It is something intrinsic.'*

Tomaź Humar came back from Nuptse after his partner Janez Jeglic had been blown off the summit, never to be found. His solo descent left him infirm for a year yet he came back. Doug Scott with 44 Himalayan expeditions, on virgin mountains or by new routes, using oxygen only once, crawled back over horrendous terrain, in great pain, from The Ogre, broken legs and all. He keeps coming back: intrinsic, ambition, madness - or just bloody minded?

Me? - 'I climb because I am'.

The Bergschrund on the Dent Blanche, in 1865. From Scrambles amongst the Alps, 1860-69, by Edward Whymper, London, 1871.

I ask is it an art, or is it a game? - For both have rules and regulations. Like football and golf, with goals, scores, penalties and handicaps. Lito Tejada-Flores in Games Climbers Play (*Mountain* 2/1969) notes,*'...a handicap system has evolved to equalize the inherent challenge and maintain the climber's feeling of achievement at a high level in each of these differing situations.'*

As the sport developed, so the goals and rules became more stringent, challenging and expansive: no fixed ropes, no aid, guide-less, alpine ascents, winter ascents, solo ascents, speed ascents, etc.. Those in the forefront of our sport provided the essence in guidance and inspiration. Teaching us to aspire to reach our own potential and to honour a joint set of ever evolving performance values, while parallel, encouraging us to see the value in our failures. In this cycle, we can inspire ourselves - each move, each decision, each thought - before, during and after.

In summary, allowing us to reach and strive for more satisfying performance levels as individuals - and hence, developing as a part of the collective soul pulsating inside the sport of mountaineering.

Climbing and mountaineering is the art of moving safely in potentially dangerous places - Like any form of art there are 'schools' and 'norms' and 'goals'. While acknowledging and accepting that they are different, similar, converging and diverging, in balance and in conflict, static and explosive, both grounded and cosmic.

The eternal question: *'Why do we climb'*, seen from today's fast-track materialistic world, is not relevant to the uninitiated and the sleeping-spectator masses. Yet, it is a useful theatrical prop for the 'spin doctors' to mystify the sport and increase their prestige.

Instead of asking, 'why?' I find it more relevant to ask how!

Mountaineering like art is a discipline. Picasso did not paint demonic creatures and one-eyed three breasted women before he was familiar with the basic tools of his art. The texture of his brushes, canvasses and the composition of his paints were as important as understanding the play of light. Nor did Shakespeare write his sonnets or Orwell his observations, or Aquinas his postulations, before they had grasped their ABC's and had studied the social interactions of mankind.

By practicing the 'basics' in their field of endeavour they progressed towards an understanding and a level of familiar proficiency, before they 'mastered' them. Then, and only then, could they express themselves totally and expand as individuals, be it in paint, music or via the written word.

So too, we, climber and mountaineer have a path to follow. One could suppose that genetic evolvement from ape to man would be an advantage, but in reality, few of us possess any real natural climbing talent. Hence, we must enter and follow the road of apprenticeship that hopefully leads to a personal and satisfying 'performance' level - and 'pure joy'.

Once the basics are mastered, we may begin the evaluation process, and as such, be free to choose our own path in the world of climbing. Be it on sunlit Mediterranean crags or solo climbing far above a potentially painful rocky floor, where a slip is judge, jury and executioner and we pay a price. Or we experience the joy of introducing children to climbing, or that of an alpine style ascent high on a new winter route on a 7000m giant. Or, or.....

It's not the **why,** it's not the **what**, it's the **how** you climb.

McDonald's Mountaineering -

The 'McDonaldisation' of mountaineering can be compared to the popular 'painting by numbers' fad in the 1960/70's. Here one could buy a line drawing showing the main features, where each section had a number that represented a given colour. By painting in the numbered spaces you eventually ended up with a Mona Lisa look-alike or a scene from Snow White and the Seven Dwarfs.

This instant piece of art could be framed and given to your half blind grandmother - who cooed at your artistic skills. As a teenager, you bathed in this adoration and came to believe that you actually had artistic talent.

So too, today's 'instant climbers' - a mirage of a lost generation.

Historically - in climbing and mountaineering, those who went before set an example. Not one necessarily needing to be surpassed. It was more an ethic of performance, the manner in which one performed. An awareness that allows us to meet the challenge the mountain presents and to embrace the forces of nature.

Today - Climbing/mountaineering has become the sport of the masses - with indoor climbing walls, tightly bolted routes, and safe guided ascents of medium difficult peaks - up iron ladders, tied into safety lines. Demanding little more than the purchase and the wearing of the necessary equipment (uniform) and that we meet up at a given place. A 'painting by numbers' phenomenon - where one does not have to strive - one only has to pay hard cash. Seeking a quick-fix alibi for one's own existence, a raison d'etre, without having to invest in oneself. A questionable path when viewed from a holistic standpoint.

How has this come about? - Commercialisation is one factor. By promoting the big names and their achievements, the equipment manufacturers, adventure travel industry, book publishers and magazines have created a 'hero'-status both for the individuals, and sadly, for the sport itself.

They sell the dream to the masses - in the form of equipment sales, travel 'expedition'-packages, all enhanced by the non-critical poorly informed, mass and specialised media who perpetuate inflated fake identities.

Everest is a good example, but not the only one, where of the 6000 or so who have stood on the summit, few have climbed the peak by another route than the two tourist ropeways. Yet, the majority of this mass have achieved a form of 'hero' status in local, national and international media and business arenas.

Climbing/mountaineering is the only sport in the world where average performance has become the norm and is acknowledged and rewarded above that of competence and excellence.

Chapter 2

Himalayan Ego-expeditions

Many expeditions we read about in the popular press, watch on TV or in films, are often led and manned by those who have not mastered the wide spectrum of disciplines found in the sport of mountaineering. Their experience usually comes via two roads. One, is to be active over a long period at a medium performance level, in reasonably familiar and safe surroundings. The other path, is to follow a climbing career purchased via intensive courses or gained under guided supervision. Both granting them the status of being, 'experienced mountaineers'.

Yet, they have rarely been involved in the decision-making process of the lead climber over a variation of challenging landscapes and under different climate conditions over time. They often lack the ability to understand and meet the demands when having responsibility for both themselves and for others - especially in 'out of their comfort zone' situations. The potentially serious consequence of their decisions, or their inability to make decisions, drives them towards, and remaining in, a cocoon of false safety.

Why go through the pain of turning theory to practice when they will never use it? For representing their interests and their interpretations of reality is the bolt, the fixed rope or the guide. They never leave that zone of complacent familiarity. Rendering them dependent on the support apparatus - be it bolt or guide.

A fast progression curve that makes them addicted to the syndrome of safety - driven by a fear of being confronted by reality - fear of failure - and the fear of being exposed.

A further warning sign are their CV's - a list of 'famous' routes that exceed their ability. This false yet strong belief in their powers is enhanced when they don their Gore-tex suits of armour, be-medalled with giant logos and dressed to receive accolades for their achievements - modern day gladiators. Honoured and feted by climbing friends and family, business associates, sponsors and the media. The circle of self-illusion is self-generating.

Training for Everest - International expedition members often meet for the first time in the bar at the Hotel de l'Annapurna or the Hotel Malla in Kathmandu. Here they are introduced to each other, tell of their backgrounds, and participate in the process of expedition planning. All contributing to the comforting illusion, that they are a part of the decision making process.

The summit of Everest is often reached that first evening, at midnight, just as the bar closes. Two days later, they fly to an airstrip near their chosen mountain, or fly directly to base camp by helicopter. Or, on the Tibetan side, they simply drive to base camp.

All the food and equipment has been organised and freighted in, kitchen and dining tents are set up, personal tents and toilet tents stand waiting. To be greeted with warm cheers by Sherpa staff after the 'arduous' approach walk. They then establish themselves in their one man roomy tents, air mattresses fully inflated and plug in their iPads.

The Icefall Doctors - Work share has changed dramatically over the years. The members of the earlier expeditions found the route through the Icefall, secured the ladders, fixed the ropes and secured the Nepali porters who carried the gear up to each camp.

A team at work.

Today, the Icefall Doctors are a commercial organisation of Sherpas, who find the easiest route (not always the safest), fixing the ropes and ladders up and through the Khumbu Icefall. Their staff, not always of the Sherpa caste, is often poorly led, poorly trained, poorly equipped, inadequately insured and poorly paid in relation to the hard work, hardship and danger they are exposed to.

This inconvenient truth, the clients are spared, as the icefall staff live in their own camp, removed from 'Sin-City': Everest Base Camp.

Then, once everything is prepared, steps cut and ladders secured, in a clear weather period, the clients are shepherded through the icefall. Their guides carry giant Thermos flasks with tea and Tang for stops along the way. For those who bother to carry a rucksack, they are filled with a puffed-up down sleeping bag, providing the ideal photo motive for sponsors, the media, and the family album.

On exiting the ice field they arrive at the first pre-established camp, to be congratulated, served a hot meal, and spend a night or two, before they descend to the comfort of base camp.

On the Tibet side, there are well stocked bars and restaurants - or they can take a quick jeep ride down to Tingri for a haircut, manicure and a full body massage.

Their main activity is not climbing, but that of moving along pre-fixed ropes and updating their commercial websites, blogs and facebooks.

Meanwhile, high altitude workers are breaking trail through the deep snow, placing the ladders, securing ropes and establishing further camps on the upper sections of the mountain. They melt snow for water, prepare the food and do the work around the camp. As the expedition progresses higher up the mountain, the high altitude workers carry extra oxygen bottles to establish a series of depots. Oxygen was originally used to assist sleeping and for the summit push. Today oxygen is used virtually all the way up, for sleeping, for the summit push and for the descent.

This work is often done under great time and economic pressure. The fine weather window for the summit bid is usually short, so everything has to be in place on time. If not, the guides and the high altitude workers do not get paid their bonus.

Pre-determined to fail - Many leaders of so called national or international expeditions cynically direct their recruiting activities towards inexperienced, incompetent and often unfit individuals. The 'hopefuls' are often tempted to make a name for themselves, or are flattered to be invited to join an expedition by a high (self) profiled climber. The final cherry, where they may have failed to gain a place in the school football team, now they will be representing their country!

Some expedition companies are confident that not all their clients will get much above Advance Base Camp, as stomach problems and altitude related sickness kick in. Or, that the client's inexperience, low level of fitness or weak psyche is not up to the enormous scale of the mountain. The long days and nights in base camp, the sound of coughs from neighbouring tents, backed by thundering avalanches and harrowing tales of people suffering from altitude and frostbite, all contribute to their erosion.

Many of these guaranteed 'pre-determined to fail' members will require neither expensive oxygen nor Sherpa support high up on the mountain, so reducing costs and increasing profits considerably.

Some hopefuls leave base camp disillusioned, never having stepped on the mountain, or only having reached Camp 1.

Back home the story changes, and they are treated to a 'hero's' welcome. To be followed by holding audio-visual shows for the local Moose Club, the company Christmas party or for the local TV station. A presentation, professionally produced, to show the polluted streets of Kathmandu, Hindu cremations, the risky landing at Lukla airstrip, the historic approach, the blessing and chanting of the Lama's, the heavily loaded yaks, and their entering base camp.

In the shadow of the killer Khumbu Icefall, the film recalls the planning, decisions and preparations, followed by days of bad weather and meetings with other expeditions. They don their crampons and we are treated to the story of their heroic filming, under extreme conditions, to document the story about their trials and tribulations as they follow in the footprints of Hillary and Tenzing.

In truth, they struggle exhausted into the higher camps, set up in advance, and wait for the guide to untie their boot laces, and be served the evening meal. The story of Hillary and Tenzing has been written and filmed a thousand times - so where is the value?

There will be tales of how many have lost their lives over the years. Their voice drops, as they recall the tragedy of losing a close climbing friend. One probably met at Sam's Bar for the first time, and bumped into along the trail. This is always a bonus, and worth a one-tear pause. Then for those who do get high, there are the endless tales of passing frozen bodies sitting in the snow. Harrowing recollections about successful and unsuccessful rescue attempts high on the mountain. Implying, but not claiming personal risk and involvement. Heroically, they play down their own suffering, by emphasizing it.

In the end they believe their own story, write yet another book on Everest, all of which helps to sustain the myth that this is mountaineering and that they are mountaineers.

Alpine Clubs and tax dodges - Some expeditions are marketed under the banner of a national or international expedition, with their national Alpine Club endorsing them. Often this is a false alibi for an extremely clever marketing strategy, business scam and tax dodge.

Big expeditions were an important part in the historical evolution of the sport of mountaineering. They have and still do play a central part in the life of many Alpine Clubs. However, mountaineering has progressed, new standards have been established and become the norms of our sport.

Unfortunately, some national and local Alpine Clubs are anachronisms and many centrally placed members are unable to accept that the use of pre-fixed ropes and pre-placed camps, the widespread use of enhancing drugs and bottled oxygen up a well-trodden peak, like Mt. Everest, by a group of medium proficient individuals, guided all the way, is irrelevant from both a national or an international mountaineering perspective.

Wake up Alpine Clubs! - Some proclaim the above methods to be the only responsible manner in which to climb.

Where the national body representing any other international sport would have debarred the individuals or team for having used unfair means, and for having put other people's lives in danger.

Their records would be deemed invalid and irrelevant.

Mark F. Twight notes, *'Managed routes, those equipped for the masses, are so convenient and safe that they may be undertaken with unknown partners picked up on the internet.'*

Messner goes deeper, *'To put on an oxygen mask is to cheat.'*

We should be allowed to ask: What is the purpose, and where is the dignity in that?

Mountaineering's Wall Street - Finance often comes from wealthy members of the expedition or from the companies they own or work for. These bogus 'national' expeditions suck out a large percent of the potential sponsorship available. From a 'mountaineering as a sport' aspect, this finance would be better spent on smaller expeditions, trying to achieve something new or worthwhile.

Unfortunately, meaningful expeditions are usually organised by mountaineers with little or no experience in the sponsor/media world. They, and the mountains they choose, are unknown to the buying public, and as such, have as much 'sex'-appeal as a deserted Welsh railway station on a stromy night.

They have nothing to offer, hence, they are unable to compete with the marketing gurus.

Question: Why do national and international companies sponsor mountaineering expeditions to big mountains that have been climbed a thousand times before, by a pre-fixed trade route?

Answer: They believe that association with an expedition to Everest, or with a high self profiled media mountaineer, will enhance the company's name, its brand and products and its reputation.

The First *'Something'* - We are witness to a cynical business model, which has turned the once proud mountain into a commodity. Being a first has high commercial value, as does being a national or an international expedition. From around 1980, the emphasis in Himalayan climbing moved away from performance in 'good style' and exploration towards that of gratification and glorification. 'Being first' entered the 'value chain' as an important 'sales tool'. It was no longer necessary to climb a new route, or gain a virgin summit. One only needed to be The First Something, by any means, at any cost.

Blind on Everest - Erik Weihenmayer, who became the first blind person to reach the summit in 2001, was ridiculed by many. Some laughed, saying, *'Everest has been climbed by the blind, the deaf and the daft... Anyone can climb Everest, people with no legs, no arms, etc...'* This may be partly true, but all ascents have a value to the individual in question - and, no matter how critical one can be to the present day climbing tourists, Everest is not a walk in the park.

What we may question in regard to 'being first' is, 'Does the ascent have any other relevance?'

In Erik's case, this was an extraordinary personal achievement. Plus, he has helped and inspired countless other handicapped people worldwide. Perhaps Erik's example will help us all to be a little more humble in our claims of heroism - and less blind to the claims of others.

Another first was the first pure Nepali Everest Expedition. This was a step in establishing the identity of Nepali national mountaineering, where they progressed from a paid support role to being independent mountaineers in their own right.

Image is everything - Many of today's expeditions have gone from exploration, challenge and adventure, to irrelevancy, repetition and mediocrity - driven by greed, egos, self-gratification and self-promoted glory.

HIMALAYAN EXPEDITION
SPECIAL OFFER

**Mount Shisha Pangma - 38 days
or 15 days Mount Cho Oyo Combination.**

- Easy snow climb of the less crowded and lowest of the world's 14 eight thousand metre, located in Tibet near Everest.
- Led by Arnold Coster, who teaches you everything you need to know. Arnold is a 4-time Everest Summiter, and this is his second Shishapangma expedition.
- Our sherpas are extremely experienced, our food is delicious and our tents and equipment are among the best.
- Climbing Shishapangma qualifies you for Everest from Tibet, Everest and Lhotse from Nepal.
- You may combine with Cho Oyu to climb Shisha in just 14 days and save 20%.
- Our last Shishapangma expedition put everyone on the summit!
- Full Service Cost (includes everything): $11,850

Book 7

RUSSIAN-ROULETTE

*'They died in harness, labouring to put their children through school,
or
to buy asthma medicine for their elderly parents.'*

Chip Brown

Chapter 1

The circus comes to Everest

I repeat - Mountaineering is the only sport in the world where mediocre performance is rewarded above that of excellence.

Most of the mountaineers in *Mount Everest (Book 5, page 133)*, were individuals or teams who performed from the middle to the very highest standards of high altitude mountaineering on Everest at a given period of time. Usually they attempted new lines, or utilised different styles of ascent -some reached the summit, some did not - but they tried.

The sport of mountaineering has been driven by the need to explore and to discover, to seek and to pass new boundaries. Standards were set, met and adhered to, then challenged and passed.

There is no climbing! - In Spring 2006, two of the world's leading mountain skiers, Tormod Granheim and Tomas Olsson, were acclimatising in preparation for an attempt to ski down the North Face of Everest. At 7070m, on the normal route, they turned around and skied down by the side of the pre-fixed ropes on the North Ridge. They skied past about 30 climbers who were on their way up.

Tormod had noticed on the way up that when one stopped for a rest that it caused a traffic jam, i.e. they all stopped and froze. Whereas he and Tomas automatically clipped out of the rope and simply climbed past the bottleneck. When they skied down he observed, *'Not one of them had an ice axe!*

The reason being, *'There is no climbing.'*

A question - *Who were these people attached to the umbilical cord - and how ready were they for Everest?*

Slovenian Tomaž Humar used to say, *'They climb the mountain in their heads one hundred times. They exhaust themselves thinking about their sponsors, the press and their self-created image. They doubt themselves; they are trapped in their own self-delusion. Exhausted, doubting and depressed before they reach base camp. Even if they reach the summit, they have failed.'*

'I think the whole attitude towards climbing Mt. Everest has become rather horrifying. At base camp there are 1,000 people and 500 tents... Just sitting around knocking back cans of beer. I don't particularly regard as mountaineering.'
Sir Edmund Hillary

Into Thin Air - Starting around 1980, commercial expeditions offered places to those who could, and would pay, and had the time.

Experience and competence were of secondary importance.

Since 1969, except 1977, there has been fatalities on Everest every year upto and including 2014. Wikipedia notes 256 dead. In 1996, ninety-eight people reached the summit, but 15 perished (8 in one day) and many others lost fingers and toes. John Krakauer's book *Into Thin Air* should be compulsory reading. As should nick Neil's *Dark Summit,* that documents yet another parody, with 11 dead in 2006. Both lessons in how not to organise and how not to partake in an expedition. Yet nothing seems to diminish the interest in the mountain - the lemmings keep coming.

Oh, and yes, they do suffer - Everest is a hard and demanding mistress. Yet to suffer or to die on a mountain does not make one a mountaineer. Nor does one become a hero just because others are suffering and dying around you. Anyone can suffer, anyone can die. It does not require a special talent.

Michael Kodas, in *High Crimes,** highlights the package holiday expedition industry where the leaders finance their own ambitions at the cost of the clients. Kodas documents poor leadership, false claims, the stealing of equipment, the sale of unsuitable oxygen apparatus, unfair bonus systems and extra costs being introduced after the climb has started - and the general disregard for other people's rights, equipment and lives.

Murder on Everest - One of those left to die was Dr. Nils Antezana, a physically fit and reasonably experienced mountaineer, who hired a professional guide, Gustavo Lisi from Argentina. Lisi, had an impressive CV, and above all, he had climbed Everest. Therefore, Nils assumed his guide had the necessary experience and competence, with the added bonus that he knew the way. Fact is, Lisi had never climbed Everest; he had failed previously and had used a photo of another climber on the summit on his own website. Why? Easy; to gain finance for his next attempt on Everest - and Nils took the bait.

* Recommended reading.

On summiting, Nils suffered from advanced hypoxia, but instead of assisting his client, Lisi spent his time filming the view and himself. Then Lisi descended, later claiming he was checking the ropes. Considering they had just been used it is highly questionable that this should take priority over his client's fast deteriorating condition.

Lisi left Nils on the summit, exhausted and suffering from life threatening oedema, with the two Sherpas, Big Dorjee and Mingma, neither of whom spoke English. They helped Nils down as far as they could, until he collapsed and they were forced to leave him. Lower down they caught up with Lisi, who was so exhausted that he had stopped and crawled into his bivi-bag. Knowing a night out at this altitude would kill Lisi, the two Sherpas forced him to continue.

Further down, with Lisi falling and staggering, Dorjee and Mingma had to leave him. On reaching camp, the other Sherpas came out to meet them, but neither indicated there was any need for assistance. They both went to their own tents, not mentioning that they had left both client and guide to fend for themselves. The other Sherpas naturally returned to their tents. Later, they heard a wailing in the night and found Lisi, about 200m from the camp, staggering, delirious and screaming in the stormy darkness.

The next day, Lisi did not inform anyone in the high camp about Nils' situation. Nor did he communicate the information to the agent in Kathmandu or to Nils' family. He simply descended, after transmitting news of his own successful 'conquest' of Everest to the world.

Everest-mania - The peak has also attracted a number of low budget climbers, often not very experienced, not very fit, and not particularly intelligent. They buy their way into an expedition organised via non-serious agents, paying for just parts of the package. They cover a percentage of the peak permit, for the use of base camp facilities, for different camps, for a given number of oxygen bottles to be deposited at specified points and for the services of a shared or their own personal Sherpa-guide. Then, as many do, they use and steal other group's equipment, including oxygen bottles, use the fixed ropes, other's tents, and depend on them for rescue and assistance.

Those western and Nepali guides and agencies who provide professional services and a solid infrastructure are justifiably concerned. It puts their groups in danger, from the stealing of equipment to adding to bottleneck situations, where people have to wait in a queue, freezing and delaying the summit push and jeopardising their descent.

Dare I ask?
'Is there a need for some form of intervention, on both sides of the border?'

If yes, then in an ideal world this would be instigated by the expedition industry itself. Unfortunately, commercial companies compete with each other, often accusing the other of taking shortcuts and of being unprofessional. The many centrally positioned individuals/board members in the Nepal Mountaineering Association (NMA), and the equivalents in China and Tibet, are the very people who organise some of the biggest and most profitable commercial expeditions - that have led to the present situation of 'death on Everest.'

The other possibility is governmental intervention and legislation. Here again, there are conflicts of interest and clear limitations. Nepali politicians have shown no interest in any serious regulation of expedition traffic. More relevant, some politicians have close family business ties with the tourist and expedition industry.

All political parties have purposely drawn out the peace process, so that the 601 members of parliament can continue to receive a teacher's yearly salary each month. While sitting in power, these criminals siphon off vast sums of foreign aid and taxes into their own projects and businesses. There were 48 highly paid Cabinet Ministers in the Maoist led coalition - yet, there were not that many ministries!

On the northern side, Chinese authorities have opened up Everest to any, and all, who can pay the fee.

Even if solutions and standards could be set and met, the question arises, 'Who should or could implement and control these?' Even more important, 'Who would police the controllers?'

High altitude workers - I refer generally to 'the Sherpas', or high altitude workers, when talking about those who work at and above Everest Base Camp (EBC). Fact is, they may come from a number of other castes, notably the Bhotia, Tamang, and Rai.

The Sherpas are but one of many castes living in Nepal. The early porters learned their trade from experienced mountaineers, and there was a natural progression from porter to high altitude porter, to a climbing support role - and to lead climbing guide. This was followed by the establishment of national climbing courses leading to a qualification. Some have taken this further to become certified international mountain guides. However, the majority of high altitude workers have very little technical mountaineering knowledge and a limited command of English.

As the majority of today's expedition members are not particularly proficient mountaineers, they are not capable of recognising weakness in security practices, nor are they qualified to instruct or control the Nepali staff. In addition, a large proportion of foreign climbers do not speak English, leading to misunderstandings, accidents and deaths.

A payment scale is based on the number and weight of loads carried up to the various camps. He who does not perform at full capacity, in all weather conditions, will not be allocated the lucrative carry sections later. Should a wrong load be delivered, or a client inconvenienced, then in all probability, they will be boycotted on future expeditions.

If the client does not summit because of weather conditions, poor leadership or bad planning by the organisation, or because of the client's lack of mental and physical fitness, lack of experience, poor acclimatising or other health problems, then the high altitude worker or guide loses the bonus.

Their working terms and conditions would not be accepted by the government nor by the unions in the countries the client climbers come from. Nor would the climbing clients themselves accept such working conditions!

Chomolungma. *Painting by: Tenzin Norbu Gurung.*

Chapter 2

Conquest of the Useless

Is mountaineering the conquest of the useless? Yes and no.
Are commercial expeditions irrelevant? Yes and no.
Their propagandists and spokesmen claim their methods to be the safest way to climb. Safest for whom, we may ask? The package-expedition industry needs to be addressed.

Dr Kenneth Kamler in his book *Doctor on Everest,* notes the tragic death of Kami in 1996, *'These things happen because we set the scene to make them happen. Once you decide that it's okay to pay people to risk their lives to help you accomplish a frivolous goal, you have to live with the consequences. Kami was young, inexperienced, careless, and he didn't clip in... we were responsible for Kami's death.*

Enticing these people to risk their lives for us is an abuse of power. We exploit them in the name of sport, offering them easy money and expedition glamour, and they don't stand a chance.'

The president of Norske Tindevegledere (NORTIND) - the Norwegian arm of the International Federation of Mountain Guides Associations (IVBV/IFMGA/UIAGM), Ola Einang, concurs. He once noted, *'We, mountaineers, know we take certain risks when practicing our sport. Yet, those we hire often trust in our ability to make the correct evaluations and to take the right decisions.'*

The list of 104 Nepali nationals who died on Everest from 1922 to 2014 reads like a liturgy. The majority come from the same district, Solu Khumbu.

Expedition work has contributed greatly to individual and family wealth and to district development. In addition, a number of Sherpas have progressed to professional guides of international standing.

Yet, the death toll speaks for itself.

Should Everest be sacrificed? - Should we encourage the mindless masses to play their games on Everest, so the rest can climb undisturbed on other mountains?

Some say yes.

I say no.

I believe that if their theatricals are exposed, then sponsors will stop sponsoring and the taxman will start taxing.

Why Everest? - Because it is the highest and she is a magnificent mountain. Yet, there are about 160 peaks in Nepal and 200 in Tibet that are still virgin. Thousands of mountains have been climbed, but only by one route. They have faces and ridges that scream to be courted. They offer incredibly beautiful and classical lines that weave their way up and through lunar like landscapes.

Not all are difficult - not all are dangerous.

I throw down the gauntlet - Travel agents, and expedition companies, there is a business opportunity here. Mr. Guide and Mr. Sherpa, you are the elite, be true to your calling. There are hundreds of mountains, in the Himalaya, still waiting to be climbed. Challenge the market, as well as yourselves and your clients, by asking them to leave the mindless rut of Everest and start mountaineering.

A menu - The Himalaya is like a menu: choose a dish, and relish the experience. One could choose lesser known and less visited peaks instead of ascending mountains that thousands have previously pulled themselves up - to descend over piles of discarded equipment, rubbish, dead bodies and human excrement.

Is it time to re-think and re-define the word 'expedition'? - A question has evolved, to ask whether the word 'expedition' should or could be examined, discussed and possibly re-defined.

One dictionary definition of the word 'expedition' is: *Noun: a sending forth or starting out on a journey, voyage, march etc. for some definite purpose as exploration or...*

This definition is pretty standard. Therefore, the majority of groups going to the mountains, including mine, would meet this definition. Be it geographical, scientific, military or personal exploration, it becomes an expedition.

If that is true, then what is the problem?

The word 'expedition' has been defiled and cheapened. We could ask, what are the intrinsic values in the word expedition? Do the words explorers, exploring, pioneers, seeking knowledge, expanding boundaries, or quest come to mind? Should this definition include entering new territory, where others have not gone before?

If not, then what?

Is it enough that it is new territory for the members themselves? If the answer is yes, then could it be time to establish guidelines on how to plan, approach and perform before and on the mountain, and how we act afterwards?

Is mediocrity in mountaineering an acceptable norm? - Should being a member of an expedition require that one possesses a minimum level of competence? That one has enough courage and integrity to acknowledge what level one has reached? Should a norm be that one contributes, has a function and is not simply passive? Should the definition of an expedition encompass that the individual and group show humbleness before, during and afterwards?

Mountaineering is a noble, yet seemingly meaningless pastime. Does and should the essential soul in mountaineering require that one respect and cherish the arena one performs in: the Mountains?

'An honest defeat is your only reward'.

These words sung by Marianne Faithful, encourage me to ask:

1. Where is the honour in paying others to face hardship in appalling conditions under great personal danger so that you may ascend Everest?

2. What is wrong in trying to be self-reliant and to climb oneself - and what is wrong with failing - with 'an honest defeat' as your only reward?

TOPOUT FACE MASKS

'...at 7500m and 8000m oxygen saturation was 73% - almost as at sea level.' - *Dr. Andrew Sutherland*

'Essential for your one shot at Everest... climb longer and use less oxygen. Topout has made the mountain lower.' - *Dr. Rob Casserley*

'...easy to use and idiot proof.' - *Dirk Stephen*

Chapter 3

The Mountain Guide

One valid argument for the acceptance and continuance of today's package holiday expedition industry is that they provide employment for the guides, agents and other workers in the industry.

The professional mountain guide has a long and honourable history. With the majority of the sport's early first ascents being done by guided clients. The guide corps of the world holds the richest bank of knowledge any sport could wish to have, providing instruction, leadership, inspiration, guidance and, where necessary, rescue resources.

Historically - the original guides in both the European Alps and in the Himalaya, were local farmers and hunters, who assisted travellers and explorers over snow covered passes and predicted the weather for them. The first known climber to hire a guide was Seigneur de Villamont in 1588, when he climbed Rochemelon, 3538m, in Italy.

The Duke of Abruzzi (1883–1933) travelled in a style similar to today's Everest expeditions. During the first ascent of Mount Saint Elias on the Alaskan coast in 1897, Abruzzi's porters carried a brass four poster bedstead to base camp. The difference was that he achieved incredible ascents, and participated in the climbing. He got to within 200 miles of the North Pole in 1899, losing two fingers in the process. In 1906, he made numerous first ascents in the Mountains of the Moon (Ruwenzori, Uganda) and confirmed that the headwaters of the Nile started from their glacial melt.

He attempted K2 (in 1909) and plotted the route that was to be followed on the first ascent. Then, with two Swiss guides, he ascended to approx 7500m (just 150m under the summit) of Chogolisa (7665m) – an altitude record not surpassed until 1922, by the second British attempt on Everest.

Guiding itself was seen as an alternative source of income for the chamois hunters and crystal gatherers of Chamonix, yet it soon turned into a viable business. In Mountaineers, the American Alpine Club and Royal Geographical Society note, '... the title passed from father to son... Most guides preferred the novice end of the market, taking well paying clients up familiar peaks'.

Not so different from today's situation in Nepal, where Sherpa guides take well paying near novice clients up familiar peaks, like Everest, with the ever present risk of orphaning Sherpa children.

The Matterhorn disaster by Gustave Doré

Nepali nationals who died when working on Mt. Everest 1922 - 2012.

Name	Accident type/Altitude	Year/Season	Expd.	Route/Place
Dorje Sherpa	Avalanche 6800m	1922 Spr	UK	N Col-N Face
Lhakpa Sherpa	Avalanche 6800m	1922 Spr	UK	N Col-N Face
Norbu Sherpa	Avalanche 6800m	1922 Spr	UK	N Col-N Face
Pasang Sherpa	Avalanche 6800m	1922 Spr	UK	N Col-N Face
Pema Sherpa	Avalanche 6800m	1922 Spr	UK	N Col-N Face
Sange Sherpa	Avalanche 6800m	1922 Spr	UK	N Col-N Face
Temba Sherpa	Avalanche 6800m	1922 Spr	UK	N Col-N Face
Shamsher Pun	AMS 5300ma	1924 Spr	UK	N Col-N Face
Man Bahadur	Exp/Frostb 5000m	1924 Spr	UK	N Col-N Face
Mingma D. Sherpa	Falling rock 6950m	1952 Aut	CH	S Col-SE Ridge
Nawang Tsh. Sherpa	Falling rock 6700m	1962 Spr	India	S Col-SE Ridge
Phu Dorje Sherpa	Crevasse 5800m	1969 Aut	Japan	S Col (recon)
Kami Tsering Sherpa	Icefall Collapse 5700m	1970 Spr	Japan	S Col
Kunga Norbu Sherpa	Icefall Collapse 5700m	1970 Spr	Japan	S Col
Nima Dorje Sherpa	Icefall Collapse 5700m	1970 Spr	Japan	S Col
Nima Norbu Sherpa	Icefall Collapse 5700m	1970 Spr	Japan	S Col
Pasang Sherpa	Icefall Collapse 5700m	1970 Spr	Japan	S Col
Tshering T. Sherpa	Icefall Collapse 5700m	1970 Spr	Japan	S Col
Kyak Tse. Sherpa	Icefall Collapse 5525m	1970 Spr	Japan	SW Face
Jangbu Sherpa	Avalanche 6900m	1973 Aut	Japan	S Col-SE Ridge
Lhakpa Sherpa	Avalanche 6400m	1974 Aut	France	Lho La-W Ridge
Nawang L. Sherpa	Avalanche 6400m	1974 Aut	France	Lho La-W Ridge
Nima W. Sherpa	Avalanche 5800m	1974 Aut	France	Lho La-W Ridge
Pemba Dorje Sherpa	Avalanche 6400m	1974 Aut	France	Lho La-W Ridge
Sanu Wangel Sherpa	Avalanche 6400m	1974 Aut	France	Lho La-W Ridge
Mingma N. Sherpa	Other 5000m	1975 Aut	UK	SW Face
Dawa Nuru Sherpa	Icefall Collapse 5800m	1978 Spr	Austria	S Col-SE Ridge
Ang Phu Sherpa	Fall 7600m	1979 Spr	SFRY	LLWR; NF-WR
Nawang Ker. Sherpa	Icefall Collapse 5700m	1980 Aut	Italy	S Col-SE Ridge
Ang Chuldin Sherpa	Avalanche 5600m	1982 Aut	Canada	S Col-SE Ridge
Dawa Dorje Sherpa	Avalanche 5600m	1982 Aut	Canada	S Col-SE Ridge
Pasang Sana Sherpa	Avalanche 5600m	1982 Aut	Canada	S Col-SE Ridge
Lhakpa Tsh.Sherpa	Non-AMS 6770m	1982 Aut	Spain	Lho La-W Ridge
Nima Dorje Sherpa	Fall 8300m	1982 Aut	Spain	Lho La-W Ridge
Pasang Temba Sherpa	Fall 8600m	1983 Aut	Japan	S Pillar-SE Ridge
Ang Rinji Sherpa	Avalanche 5600m	1984 Spr	India	S Col-SE Ridge
Jangbir Rai	AMS 4900mae	1984 Spr	India	S Col-SE Ridge
Ang Dorje Sherpa	Fall 8400m	1984 Aut	Nepal	S Col-SE Ridge
Yogendra Bdr. Thapa	Fall 8400m	1984 Aut	Nepal	S Col-SE Ridge
Phu Tashi Sherpa	Unknown	1986 Aut	UK	NE Ridge
Gyalu Sherpa	Icefall Collapse 5700m	1986 Aut	CH	S Col-SE Ridge
Dawa Norbu Sherpa	Avalanche 6600m	1986 Aut	USA	N Col-N Ridge

Name	Accident type/Altitude	Year/Season	Expd.	Route/Place
Tsuttin Dorji Sherpa	Fall 7500m	1986 Win	S Korea	SW Face (CB.R.)
Mangal S. Tamang	Avalanche 5800m	1987 Aut	UK	NE Ridge
Narayan Kr. Shrestha	Avalanche 7270m	1988 Aut	Spain	Lho La-W Ridge
Lhakpa Sonam Sherpa	Fall 8500m	1988 Aut	France	S Col-SE Ridge
Pasang Temba Sherpa	Fall 8700m	1988 Aut	France	S Col-SE Ridge
Lhakpa Dorje Sherpa	Fall 8700m	1988 Win	Belgium	S Col-SE Ridge
Phu Dorje Sherpa	Fall 8600m	1989 Spr	USA	S Col-SE Ridge
Ang Phinjo Sherpa	AMS 6500m	1989 Win	S Korea	S Col-SE Ridge
Badri Nath	Avalanche 6900m	1990 Aut	Spain	N Col-N Ridge
Ang Sona Sherpa	Avalanche 6900m	1990 Aut	Spain	N Col-N Ridge
Subba Singh Tamang	Non-AMS 5350m	1992 Spr	Spain	S Col-SE Ridge
Nun Thapa	Non-AMS 4270m	1992 Aut	Nepal	S Col-SE Ridge
Ang Gyalzen Sherpa	Other	1992 Aut	France	S Col-SE Ridge
Ang Tshering Sherpa	Crevasse 6300m	1992 Win	Spain	S Col-SE Ridge
Ms. Pasang L. Sherpa	Exhaustion 8750m	1993 Spr	Nepal	S Col-SE Ridge
Sonam Tsh. Sherpa	Fall 8750m	1993 Spr	Nepal	S Col-SE Ridge
Prem Thapa	AMS 5150m	1994 Spr	Italy	N Face (G.C.)
Mingma N. Sherpa	Avalanche 6750m	1994 Aut	Norway	N Rib-W Ridge
Kami Rita Sherpa	Fall 7100m	1995 Spr	USA	S Col-SE Ridge
Lhakpa Nuru Sherpa	Avalanche 6900m	1995 Aut	S Korea	NE Ridge
Jangbu Sherpa	Fall 8800m	1995 Aut	S Korea	N Col-N Ridge
Nawang Dorje Sherpa	AMS 6450m	1996 Spr	USA	S Col-SE Ridge
Dawa Sherpa	Avalanche 7800m	1996 Aut	S Korea	S Col-SE Ridge
Lobsang J. Sherpa	Avalanche 7800m	1996 Aut	Japan	S Col-SE Ridge
Nima Rinji Sherpa	Fall 7600m	1997 Spr	MY	S Col-SE Ridge
Mingmar Tamang	Fall 8500m	1997 Spr	S Korea	N Col-N Ridge
Tenzing Nuru Sherpa	Fall 8800m	1997 Spr	USA	S Col-SE Ridge
Babu Chiri Sherpa	Crevasse 6400m	2001 Spr	USA	S Col-SE Ridge
Karma Gyalzen Sherpa	AMS 6000m	2003 Spr	USA	S Col-SE Ridge
Bhim Bdr. Gurung	Crevasse 5900m	2003 Spr	Nepal	S Col-SE Ridge
Tuk Bdr Thapa Magar	AMS 7000m	2006 Spr	NZ	N Col-NE Ridge
Dawa Temba Sherpa	Icefall Collapse 5800m	2006 Spr	USA	S Col-SE Ridge
Lhakpa Tsh. Sherpa	Icefall Collapse 5800m	2006 Spr	USA	S Col-SE Ridge
Tenzing Phinzo Sherpa	Icefall Collapse 5800m	2006 Spr	USA	S Col-SE Ridge
Dawa Sherpa	Falling rock 7200m	2007 Spr	Italy	S Col-SE Ridge
Lhakpa Nuru Sherpa	Avalanche 5700m	2009 Spr	Nepal	S Col-SE Ridge
Kazi Lama Sherpa	Non-AMS 5300m	2009 Spr	USA	S Col-SE Ridge
Shailendra Upadhyay	AMS 5600m	2011 Spr	Nepal	S Col-SE Ridge
Karsang N. Sherpa	Non-AMS 5350m	2012 Spr	Iran	S Col-SE Ridge
Namgyal Tsh. Sherpa	Crevasse 6200m	2012 Spr	Canada	S Col-SE Ridge
Dawa Tenzing Sherpa	Non-AMS 6200m	2012 Spr	NZ	S Col-SE Ridge

Ref: Himalayan Database

Chapter 4

'We do not like to die on the snow.'
- Lhakpa Bhotia (4 time Everest summiteer).

Yama - The Lord of Death, holds the wheel of life in his mouth and embraces it with his claws, symbolising the inevitability of death, samsara and impermanence.

The Hungry Ghost - Buddhism elevates the practice of unselfishness and denounces materialistic ego-driven values such as money, fame and power. For the Sherpa to enter the realm of the Gods is to debase their sanctuary. Hence, at base camp, the ceremonial burning of a high altitude juniper-type plant, called dhupi, and prayers occur each day. To die above the snowline can result in their spirit reappearing as a Preta, the hungry ghost. A supernatural being, with a giant stomach, that symbolises perpetual craving for food and drink, and a small head and thin neck, which makes it agonising for them to eat When the sun shines, they freeze, and in the moonlight they burn.

Russian-roulette

To the best of my knowledge 116 Nepalese High Altitude Workers have died on Everest between 1922 and 2015.

The workers on Everest are exposed to continual and real danger for 70% of the expedition's total time frame. The climbing-tourists for about 20% of the same time frame. The cards are stacked..

Deaths in 2013

Namgyal Sherpa	Lobsang Sherpa
DaRita Sherpa	Mingma Sherpa

Deaths in 2014

Mingma Tenzing Sherpa	Mingma Nuru Sherpa
Dorji Sherpa	Ang Chhiri Sherpa
Nima Sherpa	Phurba Ongyal Sherpa
Lhakpa Tenzing Sherpa	Chhiring Ongchu Sherpa
Dorjee Khatri	Dorjee Sherpa
Phur Temba Sherpa	Pasang Karma Sherpa
Asman Tamang	Ang Kaji Sherpa
Ash Bahadur Gurung	Pemba Tenji Sherpa
Tenzing Chhmotar Sherpa	

Deaths in 2015 (earthquake)

Ang Kaji Sherpa	Lakpa Chhiring Sherpa
Shiva Kumar Shrestha	Pema Hissi Sherpa
Dawa Tsering Sherpa	Chhimi Dawa Sherpa
Pemba Sherpa	Milan Rai
Pasang Temba Sherpa	Tengien Bhote
Krishna Kumar Rai	Jangbu Sherpa

My question:

'If a worker from Nepal dies on Everest - Is it an accident or is it manslaughter?'

Chapter 5

Work Apartheid

The majority of early Himalayan mountaineers were experienced, competent and acclimatised.
They sought and usually found the best and safest route.
They fixed the ropes and ladders and they established the camps.
They secured their porters over the most dangerous sections.
They climbed as a team.

Many of today's Everest Tourists:

1. Are inexperienced.
2. Are incompetent.
3. Are physically and mentally unprepared for living at altitude.
4. Are ego driven, lack empathy and avoid self-evaluation.
5. Ignore the signals the mountains give and disregard sound advice.
6. Manipulate the non-critical mass and specialised media by presenting their form of mountaineering as the accepted norm, and promoting it as the only responsible form.

There are three main arguments used by the commercial expedition industry to justify their operational methods and their existance.

All three being economic:

1. Much needed money for the poor: Sherpas/staff
2. Important revenue to the country
3. Income to agents, their staff and the industry: hotels, lodges, airlines, etc.

Work Apartheid - These three crutches are used by an industry driven by ignorance, arrogance, greed and egoism. Convincing crutches utilised over the last thirty years to allow one small group, the agents, guides and politicians, to benefit materialistically, unduly out of proportion, from the work and suffering of others, those being the porters and high altitude workers.

I joust all three crutches, as they are transparent shields for the continuance of a cleverly disguised Apartheid regime.

Alibi 1 - Needed income and benefits for Sherpas, high-altitude staff and porters.

Correct. Yet emotionally misrepresentative.

Reality: High altitude workers carry heavy loads over difficult and dangerous terrain, over long periods of time, often in weather conditions experienced mountaineers would deem unacceptable, for a poor economic return in relation to fees paid into the system.

A. They operate under tight time schedules to ensure the route is constructed and all equipment is in place before the clients reach the mountain. Receive a low-base wage and a bonus system that 'forces' high 'production' levels. If loads are not delivered to the right camp at the right time they lose bonus payments, miss out on future lucrative carries, and can be boycotted from future expeditions

B. They receive poor basic training, as individuals and as a team, in the use of equipment, for climbing, descending, search and rescue.

C. They have a low level of knowledge to evaluate avalanche and weather conditions, or to evaluate their client's competence and physical/emotional performance levels. They have no authority to question the leader's decisions or the client's wishes.

D. They are poorly led.
  i. Nepali agents are rarely in the field - and if they are, they concentrate on meeting the client's needs.
  ii. Foreign expedition leaders are usually experienced (not always). They often have a double-role: 1. Salesman: in the recruitment of clients. 2. Official Leader, yet often 'demoted', so their main function is to ensure the **product sold** and the actual **service provided** match.
  iii. Sirdar. Leader, often from the Sherpa caste, possessing wide basic experience, but often only medium proficient in the craft of mountaineering. Position of high status

and authority, yet often lacking an understanding of responsible 'cause and effect' leadership. Enforcing decisions to meet the product offered, the paying client's expectations, and for their own economic returns.
    iv. Western H.A. Guide. Usually experienced, but not always in H.A. climbing. Main motive is often 'to climb Everest themselves' - working free or at low cost.
    v. Nepali H.A. Guide. Usually defers to all the above, and to the client. Often possesses a medium level of proficiency - learnt in the field, from others with low formal training, i.e. older Sirdars, and/or via medium-level courses. A growing number have 'national' certification, some hold international certification. They accompany the client on the mountain; setting up tents, melting water, cooking food, ensuring oxygen bottles are in place and function, putting on crampons etc., and clipping clients in and out of ropes - pushing them to the top, and ferrying them back down.
E. Evaluations and decisions are often taken from a position of non-equal risk sharing, such as the **2014 Spring avalanche.**

Early one morning, over forty Nepali high altitude workers were exposed to objective danger in an area of obvious high risk. Many were injured and sixteen died. Fathers and sons, brothers and uncles. The ensuing rescue/body evacuation via long line haul by helicopter over the fall area, with rising temperatures, causing dangerous updraft, placed the rescue team under great risk.

Not one western leader/guide was in the Khumbu Icefall that morning, to guide, evaluate, and to stop the bottleneck build-up.

Breaking every rule in the book.

The majority were untrained high altitude workers ferrying food and equipment to the higher camps. They were delayed at one point and the late starters kept arriving. Once the bottleneck cleared, they began to move simultaneously.

Western paying clients accompanied by a proficient guide, would have been stopped, the situation evaluated, and in all likelihood they would have descended to a safe position, and waited until the bottleneck was cleared.

Where were the leaders and where were the guides?
Michael Horst, Kurt Hunter, Joe Kluberton, Caroline Blaikie, Mike Robert, Russel Brice, Eric Simonsen and all the other responsible leaders and guides?

They were safe in lodges or in their tents, along with their clients - drinking *caffé latté* as the tragedy unfolded.

They need to be named - and shamed!
If this had happened in the USA or the European Alps, an inquiry would have followed, with the responsible guides losing their carnet, with possible legal consequences.

The Ministry of Culture, Tourism and Civil Aviation formed an investigation board comprising of representatives from the Nepal Tourism Board (NTB), Nepal Mountaineering Association (NMA), Trekking Agents Association of Nepal (TAAN) and the Expedition Operators Association of Nepal (EOAN), amongst others.

The very people and organisations that have created, managed, allowed, promoted and benefitted from this state of affairs over a long period of time. An industrial cartel that has little to be proud off.

F. Communication. Nepali staff usually have a poor command of English, as do many of the clients. A situation that has led to misunderstandings, with serious consequences.
G. Present equipment package and insurance coverage are inadequate.

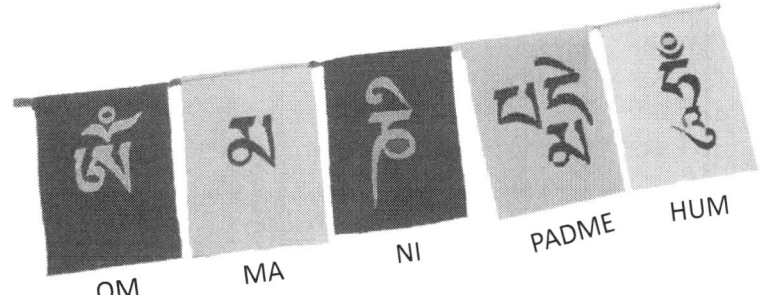

Alibi 2 - Revenue for Nepal - Peak Permit Fees - Important for the GDP of the country and major source of income for the Nepal Mountaineering Association.

Nepal has one of the most corrupt governments in Asia - billions in foreign aid, tax and revenues disappear each year. The Kathmandu airport is rated the 3rd worst in the world for service, yet substantial taxes are collected from visitors and airlines each year. The landing strip is in serious need of upgrading, as are most service and safety entities. Kathmandu is the 4th most polluted capital in the world, and the country's roads are daily death traps. The hospitals and schools are a tragic story. The government is run by hardened criminals.

Trekking (TIM's) and peak permit fees disappear into a black hole.

We are witness to a struggling Nepali Mountaineering Association, often led by the very people who run and own the expedition and trekking companies. A paralysed Tourist Department racked by nepotism, inefficiency and widespread corruption.

A recent example of years of neglect by both these organisations is the Autumn 2014 trekking disaster, where over 40 died. Where are the porter shelters, weather and avalanche warnings, communication systems and qualified guides? All supposedly financed by TIM's.

Peak permit income, although large when viewed in local currency, is mini-marginal in relation to GDP.

One must be allowed to ask, *'Where has the money gone, over all these years?'*

And to be allowed to question the answers.

Alibi 3 - Nepali travel/expedition agents provide much needed income/work.

True - Approxmately fifteen million dollars ancillary revenue. However, an out of proportion percentage of trekking and expedition revenue go to foreign agents/leaders, who organise expensive package holidays under the false flag of *national* or *international expeditions.* An emotional 'flag' that conventently hides tax scams, in their home countries.

Many Nepali expedition agents are among the richest of the rich in the travel industry - owning travel companies, lodges, hotels, equipment shops, airlines and helicopter rescue services. They have the largest houses, finest cars, invest overseas and send their children to the best schools and universities.

This sudden wealth is not all fair gain, with a proportion of agency fees being paid into offshore bank accounts to avoid tax in Nepal.

Chapter 6

An incomplete menu for suggested change

1. Legislate foreign and Nepali expedition agencies, by an independent board comprising of non-industry stakeholders (with mountaineering experience). Raise and control their accountability, financial guarantees and competence.
2. Mandatory international mountain guide certification for foreign guide leader, and/or for the group's Nepali Sirdar, or the main Nepali guide on the mountain.
3. National mountain guide certification, from the home country, for supporting foreign guides.
4. Nepali national mountain guide certification for Nepali support guides. Regular up-grade course/certification.
5. National 'High Altitude Worker' certification, to represent the work to be undertaken, for all who work above base camp
6. Qualified rescue service, in place, with relevant equipment. Clear helicopter rescue regulations - payment levels.
7. Satellite phone import procedures should be simplified, permit costs waived, to stop 'pirate' use and hence confusion
8. Enquiry into present 'Helicopter-insurance-scam' - where agents/guides get commissions, leading to unecessary use - e.g. flying people back to Kathmandu, instead of them walking, resulting in increased costs and higher premiums.
9. National weather report services to agents and expeditions.
10. Relevant quality personal equipment for all workers - including personal avalanche detectors (ref. new Recco search system.
11. Liaison Officer role to be evaluated, re-defined, with a higher level of competence/authority/accountability.
12. Insurance in the individual workers name (group insurance leads to misunderstandings). Covering rescue and repatriation and all medical costs. Realistic compensation levels for future lost income - from disability or death.
13. A clearer and fairer wage system - with standardisation of the bonus system. Pension scheme, where both workers and agents contribute an agreed percentage.

14. A verifiable and relevant climbing CV, for all clients.
15. Acclimatisation. A serious weakness, where members fly to Lukla at 2700m and reach base camp in 7 days. For Everest, Nuptse, Lhotse or Pumori the groups could go via Rowaling, Hinku Valley or over the three passes from Makalu. And/or climb a 6000 peak. This allows for realistic acclimatisation, for members to train together, and for the leader to evaluate each member before reaching Everest BC.
16. Reduce number of expedition permits to a given route on a given mountain, with a limit to the number of members.
17. Increase peak fee on popular mountains - e.g. Everest.
18. Reduce fees on less popular peaks, even granting free-permits, so spreading work/income to less developed areas.
19. National Alpine Clubs should not grant 'national expedition' status to package holiday groups, climbing by the tourist route.
20. The Himalayan Database - should only record 1st ascents of new peaks, new routes and ones of historical interest e.g. 1st national, solo, alpine, winter, and female. Repeats of standard routes by commercial groups should be deemed irrelevant by this once prestigious database.
21. Outdoor equipment brands should stop sponsoring commercial expeditions, or individuals, to popular 7000 and 8000 metre peaks, via the standard route.
22. Specialised media should reduce coverage and 'hype' around such trips.
23. National tax authorities, in Nepal and overseas, could check the financial flow of past and present expeditions - to reveal where the 'black' money went.
24. Sacralidge! Equiptment lifts by helicopter to avoid The Icefall. Alternatively: Let the climbers build the rope road! Originally the safest route up the Icefall was chosen by experienced mountaineers, assisted by local Sherpas/workers (the latter doing the heavy portage). They shared risks. As expeditions developed into 'all-inclusive package-holidays', The Icefall Doctors 'service' emerged. Owned/run by a few individuals, the actual dangerous and heavy work was done by poorly paid, poorly led and poorly equipped 'workers'. Not being experienced climbers, The Icefall Doctors chose the 'easiest' route 190- not always the safest. Once the 'road' and camps were established, along came the better paid/qualified guides and their clients: Instant Everest.

Chapter 7

'The world is a dangerous place, not because of those who do evil, but because of those who look on and do nothing.' - Albert Einstein

We have allowed the business and media moguls to define and set the premise for the sport of mountaineering and allowed them to be the purveyors of the truth.

They are not.

We have an individual and a collective responsibility, and as Einstein points out, the danger is when we look on and do nothing.

Remember that, *'He who writes history makes history.'*

They have taken the climber out of climbing and removed mountaineering from the mountain. Turning the mountain into a product to be bartered and sold to the highest bidders.

Everest by the trade route is the geriatric golf course of mountaineering

The sport of mountaineering should not be kept in the Dark Ages of package holiday expeditions.

Nick Bullock's stand, *'Consumers are dragging the mountains down to their own level with oxygen and ladders. It's time to stop pretending that standards haven't progressed and start climbing with integrity.'*

In the end, if 'their story' is repeated often enough then it becomes the 'historical truth' and as such it is no longer questioned.

Am I judgemental?

Absolutely! No doubt about it.

The other side has manipulated the media for the last thirty years to promote its version of the truth.

I do not ask nor expect all will agree.

My stand:
The sport of mountaineering has innate values that should be valued in their own right - understood, practiced, honoured and protected.

For those of you who have reached this far, I acknowledge you're staying power. Take this power to the mountains, be they the rolling hills of England, the giant rock walls of Yosemite, or the Himalaya - wherever.

Understand them, honour them, protect them, and above all enjoy them!

Namaste! - David D
Penguin Bar,
Moonlight Hotel, Thamel 2016

APPENDIX

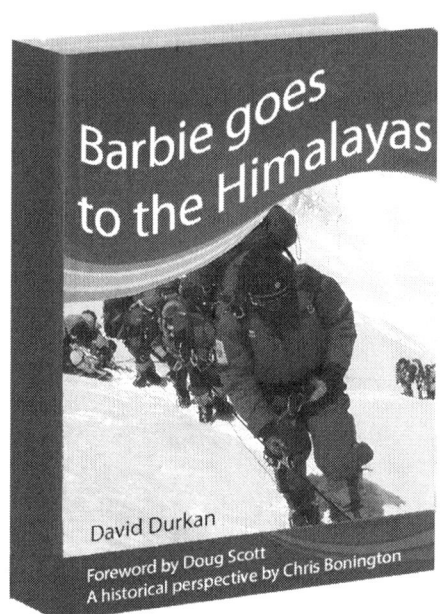

Notes and acknowledgements

The first edition (2012) started its life as *'Barbie goes to the Himalayas'*, but on learning that the name Barbie had been registered for products as diverse as toothpaste to furniture prudence prevailed. The publicity of a multi billion dollar corporation suing might be good marketing, but who wants to spend days in court? Changing the title happened without reflection, so I sincerely apologise to any penguins who feel slighted by being associated to Everest climbing tourists. I have great respect for penguins - and am concerned about their future.

I express gratitude to many, here goes: My family of origin, especially Mum and Dad, big sister Anne, as well as thousands I have met along life's road. Notably Auntie Jane and Uncle Eddy, Miss Hope, Dig, Noble Stibolt, Ken Wilson, Astri Dragni, Osho, Eric Rifkin, Ram Bahadur, Dhan Kumar, Lakpa, Reidar and Temba, Ian Wall, Sølvi Nilsen, Robert Jensen and Carl Benny.

Keith Robson and Marius M have had the difficult role of being my 'Brave Heart', over the years. A thanks to Doug Scott for his 'to the jugular' Foreword. He, along with Chris Bonington and Nils Faalund, has been supportive, inspirational and critically constructive of my endevours over the years. A special hug to Mike Tombs for his equally 'to the point' cover illustration, and thanks that he turned *Barbie* to *Penguins* with just 24 hours' notice.

Thanks to Reinhold Messner, who simply said 'yes' to the use of both his words and his name. To Michael Kodas (High Crimes) who gave a 'go ahead'. I have borrowed, unashamedly, from the National Geographic archives, from Pilgrim's *Everest* series, and from George Craig's book, *Everest,* without their permission. Seeking forgiveness here, with promises of buckets of beer to George, should our paths cross. All other quotes are poached from both the living and the dead, from books, articles or picked up in dance bars and prison cells.

Not all source references are given - a cardinal sin, for which I alone am guilty.

Photographers are credited by name, and/or by initials: GR: Gaston Rebuffat, GP: Gerda Pauler, JK: Dr. Johan Kofstad (who treated my frostbite, and saved my fingers), JH: Johnny Haglund, MI: Marcel Ichac, SH: Steve Helmore, TJ: Torgeir Kjus, P: Private - and DD being me. Historical photos are from both The Royal Geographical Society (RGS) library and from *Regards vers L'Annapurna* - B. Arthaud (et Federation Francaise De La Montagne - 1951).

Thanks to Zdeněk Thoma and to Leo Dickinson for permission to use their photos, generous gifts indeed, as are the two paintings of Chomolungma by Dolpo master Tenzing Norbu Lama. Special thanks to Ralf Dujmovits, for free use of his classic photo that tells the story far better than my 60,000 words.

Thanks to Gerda Pauler, who helped guide the first edition from start to finish. Thanks to Vanessa Knowles, Sian Pritchard-Jones, Helen Eke, Michelle Bostick and Annie, all who attempted to correct my grammar and spelling. Thanks to Rodolphe Popier, at the Himalayan Database. He, like all those noted above, is totally innocent of any errors - all of which I have added afterwards.

Quotes on page 170 (Humar), 196 (Hillary) and 197 (Norgay) are composed from various discussions, and later correspondence; as such, they are not verbatim. Other quotes used do not imply that the person in question shares all or any of my standpoints.

Thanks to Sandesh for the layout, and for not losing his cool, and to Uttam for providing a peaceful 'watering hole' at Hotel Moonlight.

Thanks to Therese and Filip, my children, for accepting me as I am, and to Beni for her deep love, tempered with a more critical eye.

Photo : Jens Riisnes

Meeting Sir Edmund Hillary at Tengboche, Khumbu, where he advised me on 'how to and how not to' build and run a school in Syangma-Tati. Something along the lines of, *'Put your money in your right pocket, and take it out of the same one, so nothing gets lost along the way. Make sure you and your Nepali partners understand and agree to who does what, why, how and when. Take nothing for granted. They often say, 'Yes', 'OK' or 'No problem' - and give you the answer they think you want.'*

He also allowed me the use of his 'office' in Kathmandu, for the start up of: **Mountain People** - *helping mountain people to help themselves*. (www.mountain-people.org). This was invaluable for securing permission for the new school, we had just built, when The Department of Education suddenly refused us permission to open it. I eventually secured a personal meeting with the Minister, who apologised for the delay. He informed me that this should not be a problem; mentioning that, *'my daughter needs a sponsor for her schooling.'* A possibility Hillary had predicted, and following his advice, I declined. The application was promptly placed in what I assumed was the 'to forget' tray.

I informed Hillary's right hand 'man' in Nepal, Liz Hawley, and she retorted, *'Silly old coot, I'll fix him.'* She phoned 'someone', resulting in the application clearly stamped 'Approved', being delivered the next week. She handed it over in a casual manner, as if twisting a Minister's arm was an everyday affair.

The school has been a success, resulting in a number of students going onto higher education, including two pilots. Twenty five years later we pulled it down and re-built it, together with the villagers.

Photo : Helge Bardseth

Interviewing Tenzing Norgay, at home in Sandvika, Norway.

As a Contributing Editor to Mountain Magazine, I interviewed Tenzing. We had three meetings, ending with a feast of arctic prawns and cold beer on my lawn. Instead of asking the usual question, *'Who reached the top first, you or Hillary?'* I tentatively touched on a point often mentioned, but never asked, *'Did you and Hillary fall out after Everest? Some believe you were jealous that Hillary received many honours, becoming rich and famous, because of Everest.'*

I held my breath. He was after all an icon in our sport, having a God-like status in both Nepal and India.

Tenzing smiled gently, *'I heard this too. No, I received all I wished for in life - Everest, family, true friends, and the Himalayan Mountaineering Institute in Darjeeling. These are my legacy. That people think because you climb one mountain together that you become friends is strange to my way of thinking. Friendship builds up over time, with deeper values than one climb, even if it is Everest.*

The British and Germans never treated us Sherpas very well. They were from the upper class, with military backgrounds, colonialists. They treated us as peons. We were given poor equipment, poor food, and low wages. Hillary was different, he was a farmer, less complex, but we were never friends. The Swiss were professional mountain guides, born and bred in the mountains, like us. They understood us better, they respected us, paying fair wages, and we were given the same equipment as they had. If I was to name one western friend, it was Raymond Lambert.'

'You laugh, you think and you question Dave's sanity! Partly biographical, he sets out to redress the balance back in favour of the high mountains.'
- Andy-psychovertical-Kirkpatrick

'An Ibsen-like odyssey - exposing the postmodern backseat drivers who strive to dominate the noble art of mountaineering.'
- Nils Faarlund

'A frontal assault on the commercialisation of mountaineering -
and how this has turned Everest Ltd.,
into a product to be bought and sold like any other comodity.'
- Kunda Dixit, Nepali Times

'Even if they reach the summit,
they have failed.'
- Tomaž Humar

Dave asks the questions
and he questions the
Thorstein Nøkleby

Move over John Cleese.
- Ian Wall

Unfair

Unfair

No crampons
No helmet
No Jummar
Load : 20-30 Kilo
One Slip = 20m Fall
Pay: US$ 22/day

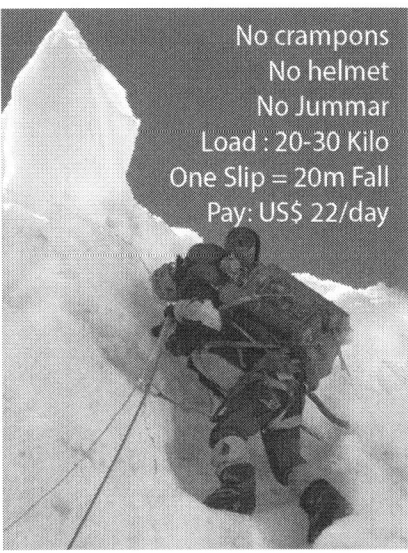

Unsightly

Summit push

Today's workers on Everest are exposed to hardship and danger for 70% of the expedition's time frame.
Climbing-tourists are only exposed to danger for about 20% of the same time frame.

Everest road construction. *Standing in line.*

Summit victory! *Rubbish left behind.*

'If a worker from Nepal dies on Everest -
Is it an accident
or
is it manslaughter?'

Printed in Great Britain
by Amazon